POLYMER SCIENCE AND TECHNOLOGY

SYNERGETICS AND FRACTAL ANALYSIS OF POLYMER COMPOSITES FILLED WITH SHORT FIBERS

Polymer Science and Technology

Additional books in this series can be found on Nova's website under the Series tab.

Additional E-books in this series can be found on Nova's website under the E-book tab.

POLYMER SCIENCE AND TECHNOLOGY

SYNERGETICS AND FRACTAL ANALYSIS OF POLYMER COMPOSITES FILLED WITH SHORT FIBERS

G. V. KOZLOV

YU. G. YANOVSKY

AND

G. E. ZAIKOV

Nova Science Publishers, Inc.

New York

For permission to use material from this book please contact us:
Telephone 631-231-7269; Fax 631-231-8175
Web Site: http://www.novapublishers.com

LIBRARY OF CONGRESS CATALOGING-IN-PUBLICATION DATA

Kozlov, G. V.
Synergetics and fractal analysis of polymer composites filled with short fibers / G.V. Kozlov, Yu.G. Yanovsky, and G.E. Zaikov.
p. cm.
Includes bibliographical references and index.
ISBN 978-1-60741-864-1 (hardcover)
1. Fiber-reinforced plastics--Analysis. 2. Polymeric composites--Mathematical models. 3. Fibrous composites--Mathematical models. I. IAnovskii, IU. G. (IUrii Grigor'evich) II. Zaikov, Gennadii Efremovich. III. Title.
TP1177.5.F5K69 2009
620.1'923--dc22

2009021049

Published by Nova Science Publishers, Inc. † New York

CONTENTS

INTRODUCTION

The interest in polymer composites is due in large degree to their many applications [1, 2]. The filling by hard disperse particulates or fibers gives to polymers desirable properties: increases stiffness, reduces thermal expansion coefficient, increases resistance to yield and fracture toughness and so on. Difficulties in research of structure-properties relationships for polymer composites are due to their structure complexity. So, on the elasticity modulus value besides composition a polymer-filler interaction influences. In other words, it is clear even intuitively that for a complete description of such structurally-complex objects as polymer composites it is necessary to account for three groups of factors, namely, polymeric matrix structure, filler structure and level of interaction between them. Up to now the researches of such types were carried out within the frameworks of continuous mechanics and thermodynamic conceptions [1, 2]. However, these conceptions application do not allow describing satisfactorily and all the more to predict such material properties. As a matter of fact, an experimental data set dictates the choice of either structure physical models for macroscopic properties description. As Ahmed and Jones showed [1], separate models in continuous mechanics were developed for the description of the experimental data specific set. Therefore a principally new approach, permitting proposing structure concrete physical representation and property description of polymer composites is required.

One of such possible approaches can be the usage of fractal analysis methods, multifractal formalism and synergetics, which have been widely practised lately [3-7]. The grounds for this are obvious. As experimental observations showed, fractal properties are inherent in the main components of polymer composites (matrix [8, 9], fibers surface [10], interfacial layers [11, 12]), i.e. they are fractal objects. This circumstance requires methods of fractal analysis application for their description [13]. Since Euclidean objects possessing translational symmetry, allow the usage of one order parameter only (the Euclidean space dimension d) for their description, then fractal objects, possessing dilatational symmetry, require the application of no than less three order parameters: Euclidean space dimension d, fractal (Hausdorff) dimension d_f and spectral (fracton) dimension d_s [13]. This requirement corresponds to the well-known Prigogine-Defay criterion in nonequilibrium thermodynamics [14] for the description of thermodynamically nonequilibrium solids, which are solid-phase polymers, requiring, as a minimum, two order parameters application. For that reason conceptions of equilibrium thermodynamics and continuous mechanics, using representations of the Euclidean geometry, widely applied for this purpose can characterize the considered materials structure only approximately and require an empirical coefficients application.

Successful application of synergetics and fractal analysis methods at polymers study [15-19], particulate-filled polymer composites and nanocomposites research [20, 21] and also the usage of late multifractal formalism [22-24] allows to expect successful spreading of this approach for structure parametrization and properties estimation of polymer composites filled with short fibers.

And let's note in conclusion that derivation of quantitative relationships structure-properties, that is the main goal of polymers physics generally, is impossible without structure quantitative identification. The structure notion is a key one in mathematics, physics, chemistry, biology and other sciences. The structure general concept satisfies to Kreber's definition: "Each system consists of elements, ordered by a definite way and connected by definite relationships. We understand system structure as elements organization mode and character of connection between them. For this it is unessential which is elements nature. Speaking of the system structure, we do not pay attention to the fact of what elements the system consists of, but consider it as relations totality only, which gives connection between system elements" [25]. To this notion the fractal analysis [3-5] and cluster model of polymers amorphous state structure [26, 27] equally satisfy. For an this, if fractal analysis represents itself mathematical calculus of high community, without allowing for specific nature of relations between elements of the structure, then cluster model is exlusively "polymeric" one, i.e. it gives structure description in terms of definitions, generally accepted in physics of polymers. In virtue of the indicated circumstances these models are the excellent complement of one another and allow to obtain the most complete quantitative description of polymer composites structure.

REFERENCES

[1] Ahmed S., Jones F.R. *J. Mater. Sci.*, 1990, v. 25 № 12, p. 4933-4942.

[2] Polymer Engineering Composites. Ed. Richardson M.O.W. London, *Applied Science Publishers LTD,* 1978, 459 p.

[3] Mandelbrot B.B. *The Fractal Geometry of Nature.* New York, W.W. Freeman and Company, 1982, 459 p.

[4] Shabetnik V.D. *The Fractal Physics. Science on World Creation.* Moscow, OAO "Tibr", 2000, 326 p.

[5] Feder F. *Fractals.* New York, Plenum Press, 1990, 248 p.

[6] Balankin A.S. *Synergetics of Deformable Body.* Moscow, Publishers of Ministry Defence SSSR, 1991, 404 p.

[7] Bobryshev A.I., Kozomazov V.N., Babin L.O., Solomatov V.I. Synergetics of Composite Materials. Lipetsk, *NPO ORIUS,* 1994, 154 p.

[8] Bagryanskii V.A., Malinovskii V.K., Novikov V.N., Pushchaeva L.M., Sokolov A.P. *Fozoka Tverdogo Tela,* 1988, v. 30, № 8, p. 2360-2366.

[9] Zemlyanov M.G., Malinovskii V.K., Novikov V.N., Parshin P.P., Sokolov A.P. *Zhurnal Eksperimental'noi i Teoreticheskoi Fiziki,* 1992, v. 101, № 1, p. 284-293.

[10] Avnir D., Farin D., Pfeifer P. Nature, 1984, v. 308, № 5959, p. 261-263.

[11] Novikov V.U., Kozlov G.V., Bur'yan O.Yu. *Mekhanika Kompozitnykh Materialov,* 2000, v. 36, № 1, p. 3-32.

[12] Kozlov G.V., Yanovskii Yu.G., Lipatov Yu.S. Mekhanika Kompozitsionnykh Materialov i Konstruktsii, 2002, v. 8, № 1, p. 111-149.

[13] Rammal R., Toulouse G. *J. Phys. Lett.* (Paris), 1983, v. 44, № 1, p. L13-L22.

[14] Nemilov S.V., Bogdanov V.N., Nikonov A.M., Smerdin S.N., Nedbay A.I., Borisov B.F. *Fizika i Khimiya Stekla,* 1987, v. 13, № 6, p. 801-809.

[15] Novikov V.U., Kozlov G.V. *Uspekhi Khimii,* 2000, v. 69, № 4, p. 378-399.

[16] Novikov V.U., Kozlov G.V. *Uspekhi Khimii,* 2000, v. 69, № 6, p. 572-599.

[17] Kozlov G.V., Novikov V.U. *Synergetics and Fractal Analysis of Cross-linked Polymers.* Moscow, Klassika, 1998, 112 p.

[18] Shogenov V.N., Kozlov G.V. Fractal Clusters in Physics-Chemistry of Polymers. Nal'chik, *Polygraphservice and T,* 2002, 268 p.

[19] Kozlov G.V., Zaikov G.E. *The Structural Stabilization of Polymers: Fractal Models. Utrecht-Boston,* Brill Academic Publishers, 2006, 345 p.

[20] Aloev V.Z., Kozlov G.V. Physics of Orientational Phenomena in Polymeric Materials. Nal'chik, *Polygraphservice and T,* 2002, 288 p.

[21] Malamatov A.Kh., Kozlov G.V., Mikitaev M.A. *Reinforcement Mechanisms of Polymer Nanocomposites.* Moscow, Publishers of Mendeleev RKhTU, 2006, 240 p.

[22] Novikov V.U., Kozitskii D.V., Deev I.S., Kobets L.P. *Materialovedenie,* 2001, № 11, p. 2-10.

[23] Novikov V.U., Kozlov G.V. Plast. *Massy,* 2004, № 4, p. 27-38.

[24] Kozlov G.V., Novikov V.U., Zaikov G.E. *J. Balkan Tribologic. Assoc.,* 2004, v. 10, № 2, p. 125-199.

[25] Ebeling W. Structurbildung bai irreversiblen Prozessen. Eine Einführung in die Theorie dissipative Structuren. Berlin, BSB B.G. *Teuber Verlag-Sgesellschaft,* 1976, 279 s.

[26] Kozlov G.V., Novikov V.U. U*spekhi Fizicheskikh Nauk,* 2001, v. 171, № 7, p. 717-764.

[27] Kozlov G.V., Zaikov G.E. Structure of the Polymer Amorphous State. Utrecht-Boston, *Brill Academic Publishers,* 2004, 465 p.

Chapter 1

The Structure of Polymer Composites Filled with Short Fibers

1.1. Synergetics of Composites Structure Formation

At present views on amorphous polymers structure are subjected to cardinal revision owing to vigorous development of synergetics and fractal analysis. As is known [1], synergetics studies universal laws of spatial structures self-organization in dynamic systems of various nature. Let's note in this connection three important aspects. Firstly, the systems adaptation process to external influence by a structures self-organization way can proceed in nonequilibrium systems only, which are solid-phase amorphous and semicrystalline polymers. Secondly, such self-organizing (dissipative) structures are dynamic ones, reacting on any change of environment. And, thirdly, the feature of dissipative structures, forming in polymers, is spatial scales (structural levels) existence, that founds its reflection in structure linear sizes hierarchy [2].

As is well known, polymer composites are heterogeneous structurally-complex solids, the main structural components of which consist of polymeric matrix, filler and interfacial regions. The synergetic behaviour implies interconnection (interaction) of the indicated structural components and the existence of feedback between them [1]. Therefore, the clarification of purely material-technical aspects of this problem is of interest, i.e. what structures are formed at using composites preparation method, how they interact and how feedback between them is realized. This interest intensifies the circumstance, that composites properties depend on their structural components characteristics combination by enough complex mode. Therefore the authors [5-8] made the study of the structure formation of carbon plastics on the basis of the thermostable aromatic polyamide phenylone S-2 filled with carbon fibers (CF) using synergetics methods [1], fractal analysis [2] and cluster model of polymers amorphous state structure [3, 4].

Before consideration of the indicated composites structure formation synergetics it is necessary to give a brief description of their preparatory method. Composites with CF mass contents 15 %, that corresponds to the filling volume degree $\varphi_f \approx 0.115$, were prepared by "dry" method, including components blending in rotating electromagnetic field [9]. For this purpose, powdery polymer, CF, and nonequiaxial ferromagnetic particles with length of 40 mm were placed in a reactor. Then the reactor was placed on the end window of the generator of the electromagnetic apparatus. Under the influence of the rotating electromagnetic field, ferromagnetic particles began to rotate, colliding with each other, that resulted to the equipartitional (chaotic) distribution of the CF in the polymer matrix. As a result of the collisions, the particles were worn down, and the products of wear became part of the composition. After blending, two methods were used to remove the ferromagnetic particles: magnetic and mechanical separation [8]. The specimens for properties studying were prepared by a method of hot pressing.

Detailed study of the dependences of carbon plastics properties number on components blending duration t in rotating electromagnetic field discovers one common feature, having statistical character: at first the periodic (ordering) behaviour, close to the sine one with period doubling and then the transition to chaotic behaviour is realized [8]. Such behaviour is typical for synergetic systems [1]. For quantitative description of the observed effect the authors [8] have calculated fractal (Hausdorff) dimension d_f of carbon plastics structure, which is the universal informant of substance structural state, according to the equation [10]:

$$d_f = (d-1)(1+\nu), \qquad (1.1)$$

where d is dimension of Euclidean space, in which a fractal is considered (it is obvious, in our case d=3), ν is Poisson's ratio, determined according to the mechanical tests results with the aid of the relationship [11]:

$$\frac{\sigma_Y}{E} = \frac{1-2\nu}{6(1+\nu)}, \qquad (1.2)$$

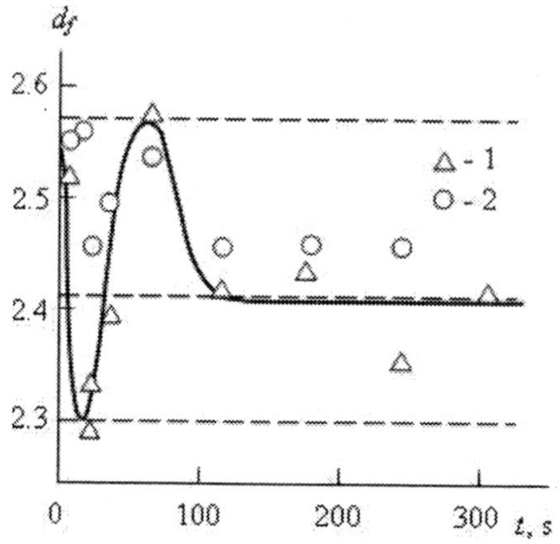

Figure 1.1. The dependence of structure fractal dimension d_f on duration t of components blending in rotating electromagnetic field for carbon plastics on the basis of phenylone, obtained with magnetic (1) and mechanical (2) separation application. Horizontal shaded lines indicate the threshold values d_f, corresponding to critical indices of percolation [8].

where σ_Y is yielding stress, E is elasticity modulus.

The dependence $d_f(t)$ for both series of the considered carbon plastics is shown in Figure 1.1. As one can see, in the range t=5-120 s the approximately sine dependence $d_f(t)$ is obtained, which at $t \geq 120$ s approaches to the constant value $d_f \approx 2.41$. As it was noted above, such type of dependence was distinctive for periodic (quasiperiodic) structures with subsequent transition of the system to chaotic behaviour. This observation indicates that the postulated above steady (chaotic) CF distribution at the described composites blending method is not achieved instantly, but only at $t \geq 120$ s, whereas at $t < 120$ s carbon plastics behaviour is controlled by periodic (quasiperiodic) or ordered structures [1].

Let's note a very important technological aspect. As follows from the data of Figure 1.1, the values d_f for the carbon plastics, obtained with magnetic separation application, are lower than the corresponding values for mechanical separation case (the average values $d_f=2.413$ and 2.529, respectively). This effect can be due to the circumstance, that in mechanical separation case in blend wear products of ferromagnetic particles remain, which can prevent from local order domains (clusters) formation and thereby increase the value d_f [3, 4].

Let's further estimate theoretically the values d_f, corresponding to minimum ($d_f^{\min}$), maximum ($d_f^{\max}$) and transition to chaotic behaviour (d_f^{ch}) on the dependence $d_f(t)$. As it is known [5], polymeric binding of carbon plastics consists of three structural components: bulk polymeric matrix includes local order domains (clusters) and loosely-packed matrix and the third structural component is interfacial regions, which parallelly with clusters are related to densely-packed structure regions (more precisely, to the regions, where molecular mobility is inhibited), whereas loosely-packed matrix presents itself chaotic component of amorphous polymers structure, consisting of chaotically tangled parts of macromolecular coils [4]. This allows to connect carbon plastics structure ordered behaviour with clusters and interfacial regions and chaotic one – with loosely-packed matrix [8].

In paper [12] applicability of the thermal cluster model for polymer composites structure description is shown. As it is known [13], percolation cluster, presenting purely geometrical construction, is a too simplified model for real amorphous polymers, possessing thermodynamically nonequilibrium structure. Therefore polymer thermal interactions, influence on the structure of the mentioned polymers; that should be understood as molecular mobility, i.e. thermal oscillations of macromolecules fragments around their quasiequilibrium positions [14]. Besides, the thermal cluster formation is studied not on solid-state component concentration scale, as in case of percolation systems [15], but on relative temperatures scale [16-18]. For thermal cluster a relative fraction of local order domains (clusters) φ_{cl}, i.e. order parameter [16, 17], is connected with glass transition T_g and testing T temperatures by the following relationship [18]:

$$\varphi_{cl} = \left(\frac{T_g - T}{T_g} \right)^{\beta_T}, \qquad (1.3)$$

where thermal cluster index β_T is not definitely equal to corresponding index of order parameter β_p in geometrical percolation models.

It also has been shown [12], that the critical indices β_p, ν_p and t_p of percolation cluster (which are equal to 0.40, 0.80 and 1.60, respectively [15]) are border values for β_T, indicating which structural component of composite defines its behaviour. At $\beta_T=\beta_p$ such component are cluster or, more precisely, percolation system network, identified with cluster network. At

$\beta_p<\beta_T<\nu_p$ composite behaviour is due to the combined influence of clusters and loosely-packed matrix. At $\beta_T=\nu_p$ a determining structural component will be a loosely-packed matrix, at $\beta_T=t_p$ – the filler particles network and at $\nu_p<\beta_T<t_p$ the combined influence of two last structural components is observed. Let's note, that under filler particles network influence the polymeric matrix-filler interfacial layers totality is implied. The average value of T_g for the considered carbon plastics is accepted equal to 530 K [8].

For $d_f=d_f^{min}$ $\beta_T=\beta_p=0.40$ and according to the equation (1.3) $\varphi_{cl}=0.74$ for $T=293$ K will be obtained. Further the value d_f^{min} can be estimated theoretically with the aid of the equation [4]:

$$d_f = 3 - 6\left(\frac{\varphi_{cl}}{SC_\infty}\right)^{1/2}, \qquad (1.4)$$

where S is macromolecule cross-sectional area, C_∞ is characteristic ratio, which is an indicator of polymer chain statistical flexibility [19]. For phenylone $S=17.6$ Å^2 [20], $C_\infty=3$ [21].

For $\beta_T=\beta_p=0.40$ according to the equation (1.4) the value $d_f^{min}=2.30$ will be obtained, that corresponds excellently to the experimentally obtained dimension (Figure 1.1). For $\beta_T=t_p=1.60$ $\varphi_{cl}=0.265$ and $d_f=d_f^{max}=2.575$, that again well corresponds to the experiment. Such correspondence is obtained and in case of transition to chaotic behaviour: $\beta_T=\nu_p=0.80$, $\varphi_{cl}=0.515$ and $d_f=d_f^{ch}=2.407$. These limiting values d_f are indicated in Figure 1.1 by horizontal shaded lines. In such treatment the period doubling reason becomes obvious – it corresponds to consecutive doubling of critical indices β_p, ν_p and t_p [8].

As it is known [1], for synergetic systems the common dependence is observed: at external parameter change the system behaviour changes from simple to chaotic one. However, a definite range of external parameter exists, in which system behaviour is ordered and periodic. The data of Figure 1.1 suppose, that such range is period $5\le t\le 120$ s and at $t>120$ s the transition to chaotic behaviour is observed. The orderliness contains in the fact, that in each time moment the system behaviour is reproduced. The doublings number a can be received from the following equation [1]:

$$Z_\infty - Z_n = \delta^{-a}, \qquad (1.5)$$

where Z_n is a guiding parameter value, at which period is doubled in n number, Z_∞ is a limiting value of this parameter, δ is Feigenbaum constant ($\delta\approx4.67$ [1]).

Since at the considered mode of carbon plastics preparation the same polymeric binding and filler are used at constant content of the latter, then it is natural to suppose, that carbon plastics structure and properties change as function t is the result of filler structure variation. Under filler structure the combination of such factors as fibers orientation, their aggregation degree and fibers distribution in polymeric matrix is implied [22, 23]. As it is known [22], the composite fracture stress value σ_f^c can be calculated according to the equation:

$$\sigma_f^c = \eta\tau_Y\left(\bar{l}/\bar{d}\right)\varphi_f + \sigma_f^m\left(1-\varphi_f\right), \tag{1.6}$$

where η is fibers orientation factor, τ_Y is matrix shear yield stress, ($\bar{l}/\bar{d}$) is ratio of average magnitudes of fibers length and diameter, σ_f^m is polymeric matrix fracture stress.

At η calculation according to the equation (1.6) the following values of including in it parameters were accepted: $\tau_Y=\sigma_Y/\sqrt{3}=133$ MPa [24], ($\bar{l}/\bar{d}$)=300 ($\bar{l}$=3 mm, $\bar{d}$=7-9 mcm), φ_f=0.115 and σ_f^m=235 MPa [25]. It is supposed, that in the first approximation ($\bar{l}/\bar{d}$)=constant. It is natural, that in preparation process some $\bar{l}$ reduction can occur [25], but it will be a monotonous t function and it will not influence qualitatively on the calculating value η.

In Figure 1.2 the dependence d_f on η for both series of the studied carbon plastics is adduced. As one can see, the linear correlation $d_f(\eta)$ is obtained, but with large scattering. Such scattering can be due to the condition σ_f^m=const choice. As follows from the graph of Figure 1.1, carbon plastics structure changes at t variation and its change should lead to σ_f^m constancy condition violation. σ_f^m estimation as function t (or d_f) allows the following equation [26]:

$$\sigma_f^m = 1.4\times10^5\left(\frac{\varphi_{cl}}{2N_A S l_0 C_\infty}\right)^{5/6}, \text{ Pa}, \tag{1.7}$$

where N_A is Avogadro number, l_0 is the main chain skeletal bond length, which is equal to 1.49 Å for phenylone [21].

In Figure 1.3 the dependence $d_f(\eta)$, where σ_f^m value was calculated according to the equation (1.7), is adduced. As one can see, in this case much better linear correlation is obtained. Then, using $Z_\infty=\eta_\infty=0.41$ (t=60 s) and $Z_n=\eta_n=0.21$ (t=120 s), one will obtain a=1.11 according to the equation (1.5), i.e. one-multiple period doubling is expected theoretically, that is confirmed experimentally (Figure 1.1).

At the first sight, the data of Figs. 1.2 and 1.3 suppose that η is a guiding parameter for carbon plastics structure. However, d_f increase at η growth means corresponding rise of relative fluctuation free volume f_g according to the equation [3]:

$$f_g \approx 8.5\times10^{-3}\left(\frac{d_f}{d-d_f}\right) \tag{1.8}$$

and hence entropy change ΔS growth [27]:

$$\Delta S \approx -3kf_g \ln f_g, \tag{1.9}$$

where k is Boltzman constant.

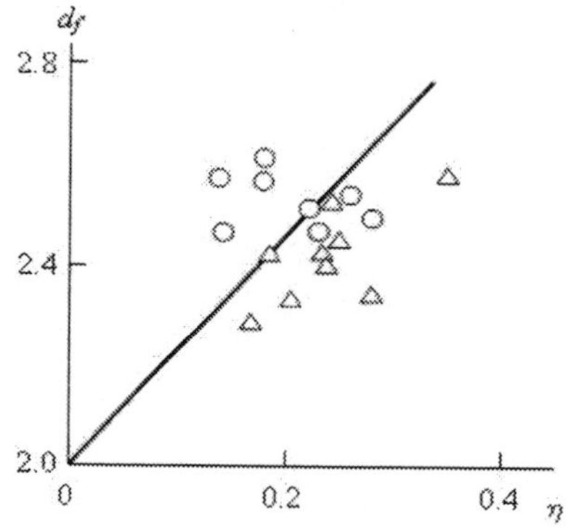

Figure 1.2. The dependence of structure fractal dimension d_f on fibers orientation factor η, calculated at σ_f^m =const, for carbon plastics on the basis of phenylone. The designations are the same as in Figure 1.1 [8].

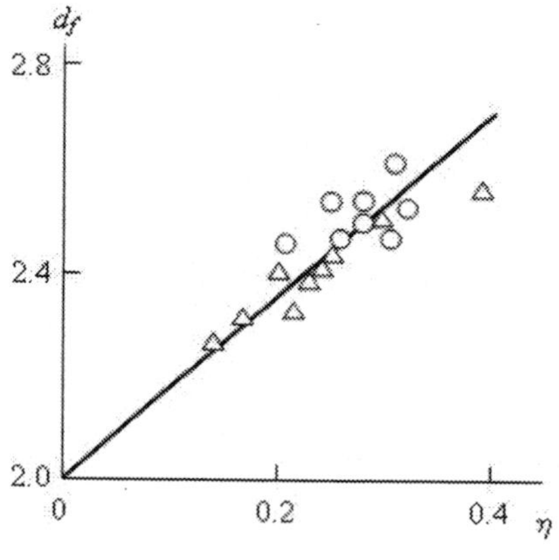

Figure 1.3. The dependence of structure fractal dimension d_f on fibers orientation factor η, calculated at σ_f^m estimation according to the equation (1.7), for carbon plastics on the basis of phenylone. The designations are the same as in Figure 1.1 [8].

The considered above relationships contradict to Klimontovich S-theorem, from which the reduction of normalized by kT entropy at η increase should follow [28]. In this case only η is the structure guiding parameter in strict physical significance of this term. The adduced in Figs. 1.2 and 1.3 dependences do not correspond to the entropy production minimum principle, i.e. they mean carbon plastics structure self-organization process absence [1]. However, as it has been noted earlier, polymer composite is heterogeneous structurally-complex solid, consequently one of its structural components can be "leading", i.e. it can be formed in the first place define other component parameters and the other – "supporting". Such interconnection is observed at polymers crystallization, when forming in the first place crystalline phase rejects in interfacial regions chain irregularities of various kinds, defining by the latter their structure. Such law availability in synergetic systems confirms the obligatory existence of feedback in them [29]. For carbon plastics structural expression of feedback finds its reflection in interconnection of the two most densely-packed structure components: clusters and interfacial regions with relative fraction φ_{if}, that is described analytically by simple relationship [5]:

$$\varphi_{cl} = 0.74 - \varphi_{if}. \tag{1.10}$$

Let's note that the constant 0.74 in the relationship (1.10) is equal to maximum relative fraction of composite densely-packed regions according to the thermal cluster conception, i.e. according to the equation (1.3).

The value φ_{if} can be calculated according to the following equation [30]:

$$\varphi_{if} = 1 - \frac{\Delta C_p^c}{\Delta C_p^p}, \tag{1.11}$$

where ΔC_p^c and ΔC_p^p are heat capacity at constant pressure jump values at glass transition temperature for composite and matrix polymer, respectively.

Self-organization conception is connected inseparably with self-government by inverse connections action way [1]. For systems with feedback nonlinear dependence of previous (η_n) and subsequent (η_{n+1}) values of governing parameter is characteristic, which can be written in Puancare equation form [1]:

$$\eta_{n+1} = \eta_n (1 - \eta_n)\lambda, \tag{1.12}$$

where λ is parameter of feedback.

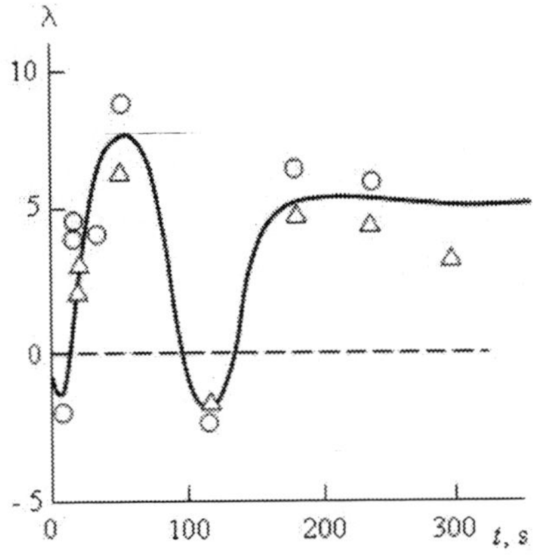

Figure 1.4. The dependence of feedback parameter λ on duration t of components blending in rotating electromagnetic field for carbon plastics on the basis of phenylone. The designations are the same as in Figure 1.1 [8].

In Figure 1.4 the dependence $\lambda(t)$ for both series of the studied carbon plastics is adduced (let's note, that $\Delta_i=\lambda^{-1}$ is system stability measure, i.e. structure of composite). As one can see, in the range t=5-60 s the fast growth λ is observed, system decreases its stability. Therefore the maximum stability of carbon plastics structure one must connect with cluster network and the minimum one – with interfacial regions. At t=120 s the negative feedback is obtained, that is criterion of transition to chaotic behaviour.

In Figure 1.5 the dependences of φ_{cl} and φ_{if} on duration t of components blending in rotating electromagnetic field for carbon plastics, obtained with magnetic separation application. At small t the clearly expressed antibatness of the curves $\varphi_{cl}(t)$ and $\varphi_{if}(t)$ is observed, corresponding to the relationship (1.10) and, consequently, feedback principle action, which violates in the range $t\approx60$ s, i.e. at system transition to chaotic behaviour [1].

It is obvious enough, that the first structural component, which will respond to the distribution character change of carbon fibers in polymeric matrix, is interfacial regions, forming by stretching and stacking of macromolecules on these fibers relatively smooth surface [31]. This means that for the considered carbon plastics the interfacial regions are the "leading" component and the "supporting" one – the cluster network (according to the relationship (1.10)). For this assumption checking the entropy change for interfacial regions ΔS^{if} was calculated according to the equation (1.9), where f_g value was determined according to the formula (1.8) and d_f – according to the equation (1.4) at the replacement φ_{cl} on φ_{if} and at the condition C_∞=9 [5]. In Figure 1.6 the dependence of normalized by kT entropy change for interfacial regions on η for both series of the studied carbon plastics is adduced. As one can see, in this case fibers orientation factor η is a governing parameter for interfacial regions – in correspondence with S-theorem η growth results to these regions normalized entropy reduction. As should be expected, the departure from general correlation is observed at $t\approx60$-120 s, i.e. at negative feedback and transition to chaotic behaviour [1].

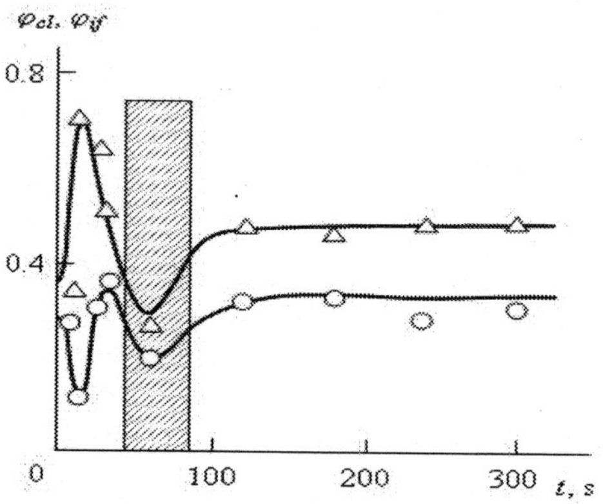

Figure 1.5. The dependence of relative fraction of clusters φ_{cl} (1) and interfacial regions φ_{if} (2) on duration t of components blending in rotating electromagnetic field for carbon plastics on the basis of phenylone, obtained with magnetic separation application. The shaded region indicates transition to chaotic behaviour [8].

In Figure 1.7 the relation between values φ_{cl} and φ_{if} is shown, which corresponds to the equation (1.10) and expresses feedback principle for the studied carbon plastics. As one can see, the data departure from general correlation at $t \approx 60$-120 s is observed again, which is due to the indicated above reasons.

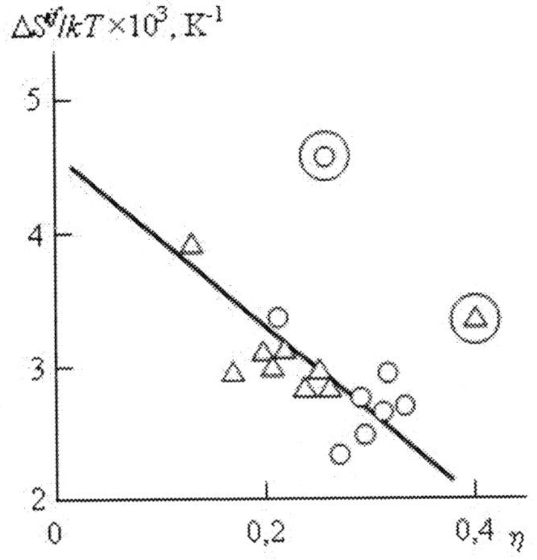

Figure 1.6. The dependence of normalized change of interfacial regions entropy $\Delta S^{if}/kT$ on fibers orientation factor η for carbon plastics on the basis of phenylone. The data, corresponding to transition to chaotic behaviour, are outlined with small circle. The designations are the same as in Figure 1.1 [8].

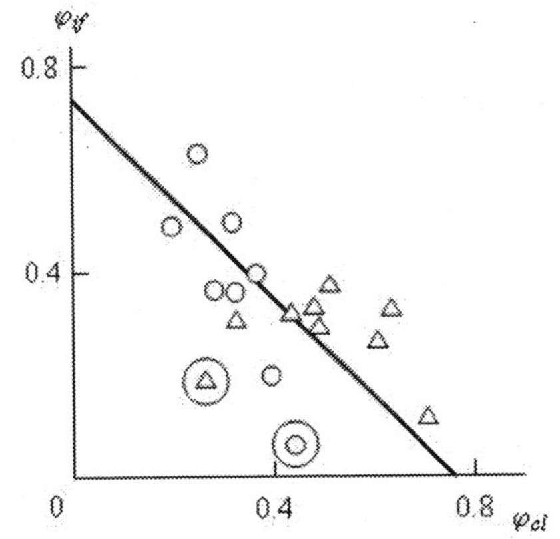

Figure 1.7. The relation between relative fractions of clusters φ_{cl} and interfacial regions φ_{if} for carbon plastics on the basis of phenylone. The straight line is drawn according to the relationship (1.10). The designations are the same as in Figs. 1.1 and 1.6 [8].

Therefore, the stated above results have shown that carbon plastics structure corresponds completely to synergetic systems main laws and characteristics both qualitatively and quantitatively. The experimentally obtained threshold values of fractal dimension and period doublings number correspond completely to theoretical calculation. The system ordered behaviour associates with structure densely-packed regions and chaotic one – with loosely-packed matrix. The fibers orientation factor in polymeric matrix is a governing parameter for interfacial regions, formation of which controls cluster formation in virtue of feedback principle action [8].

The authors of paper [32] used another synergetic treatment of carbon plastics structure formation with application of nonequiaxial ferromagnetic particles of length l_f=20, 40 and 70 mm. The discrete-wave model of amorphous polymers supramolecular structure is assumed as a basis of this treatment, which accounts for the unity of discrete and wave properties of substance in the condensed state [33]. In this case "scenario" of formation and development of structural levels hierarchy can be presented with the aid of iterative process [33]:

$$l_k = \langle a \rangle B_\lambda^k \text{, where } k=0, 1, 2, \ldots \tag{1.13}$$

In the equation (1.13) l_k is characteristic spatial scale of structural changes, B_λ is discrete-wave criterion, which is equal to ~ 2.61 [31], k is number of structural hierarchy sublevel.

As it was shown earlier [2], for k=0 $l_k=\langle a\rangle=l_0$, where l_0 was the main chain skeletal bond length, which is equal to 1.47 Å for phenylone [21]. Within the frameworks of hierarchical model of amorphous polymers structure the first three linear scales are written as follows [2]:

$$\Lambda_0 = l_0; \Lambda_1 = l_0 C_\infty = l_{st}; \Lambda_2 = l_0 C_\infty^2, \tag{1.14}$$

where l_{st} is statistical segment length [34].

From the comparison of the equations (1.13) and (1.14) qualitative similarity of the concepts [2] and [33] follows, but the principal distinction is constant B_λ replacement on molecular variable C_∞ (in general case $C_\infty \neq B_\lambda$). The fractal dimension d_f of carbon plastics structure can be estimated with the aid of the equation [35]:

$$I = 2\times10^{-6}\left(d_f - 2.5\right), \tag{1.15}$$

where I is linear wear intensity in friction process.

Let's note, that the calculated according to the equation (1.15) d_f values describe carbon plastics structure at average testing temperature T_{av}=400 K [35] and for l_f=20 mm [32]. Supposing for this temperature C_∞=2, the value d_f can be calculated according to the following equation [2]:

$$C_\infty = \frac{2d_f}{d(d-1)(d-d_f)} + \frac{4}{3}. \tag{1.16}$$

The dependence $d_f(t)$, where d_f values were calculated according to the equation (1.15), at l_f=20 mm, was shown in Figure 1.8. As one can see, its shape is identical to the adduced one in Figure 1.1 and higher d_f values are due to the increased testing temperature. Besides, in Figure 1.8 the theoretical values d_f, calculated according to the equation (1.16), are shown by horizontal shaded lines, which correspond to k=3-8. As one can see, the calculated according to the equation (1.15) d_f values correspond well to discrete d_f magnitudes according to the equations (1.14) and (1.16). The values d_f, answering to nonoccupied levels k=5, 6 and 7, are observed for other values of ferromagnetic particles length l_f [32].

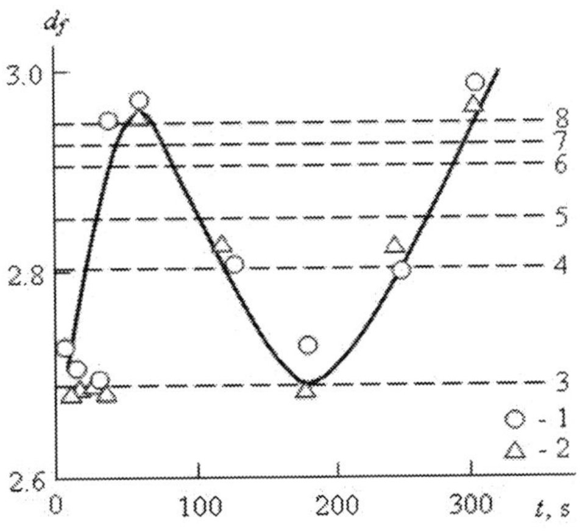

Figure 1.8. The dependence of structure fractal dimension d_f on duration t of components blending in rotating electromagnetic field for carbon plastics on the basis of phenylone, obtained with mechanical (1) and magnetic (2) separation application. The horizontal shaded lines indicate k discrete levels for d_f value [32].

To trace carbon plastics structure formation cyclicity is possible according to the data of Figure 1.9, where the dependences $k(t)$ for l_f three values (20, 40 and 70 mm) are adduced. As one can see, in all three cases the sine dependence $k(t)$ is observed, but the first maximum reaching time t_{max} at k=8 decreases systematically at l_f growth. In Figure 1.10 the dependence t_{max} on l_f reciprocal value is adduced, which turns out to be linear and passing through coordinates origin.

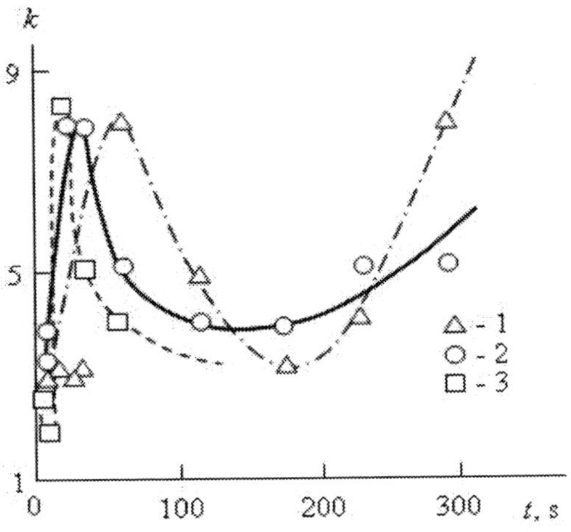

Figure 1.9. The dependences of structural hierarchy level k on duration t of components blending in rotating electromagnetic field for carbon plastics on the basis of phenylone at following lengths of ferromagnetic particles l_f: 20 (1), 40 (2) and 70 (3) mm [32].

As a matter of fact, discrete-wave model of carbon plastics structure formation can be considered within the frameworks of multifractal formalism [36], where structure fractal dimension d_f depends on measurement scale Λ_1 or l_{st}, determined according to the equation (1.14). In Figure 1.11 such dependence $d_f(l_{st})$ for l_f=20 mm is adduced, from which it follows, that it has typical for similar diagrams "bell-like" shape. Let's note, that this dependence is a generalized one, i.e. d_f values and corresponding to them l_{st} magnitudes are obtained for different carbon plastics [32].

As it should be expected, sine change of structure characteristic d_f defines similar change of these materials properties. In Figure 1.12 the comparison of schematic dependences $d_f(t)$ according to the data of Figure 1.8 and yield stress σ_Y on t, is adduced from which their analogy follows.

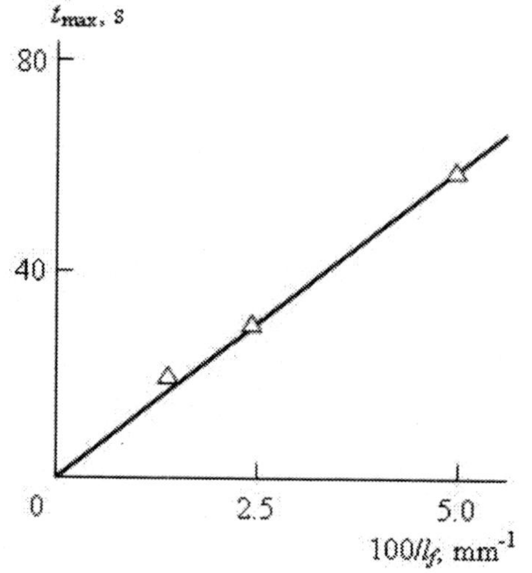

Figure 1.10. The dependence of the first maximum on the curve $k(t)$ (Figure 1.9) reaching time t_{max} on reciprocal value of ferromagnetic particles length l_f [32].

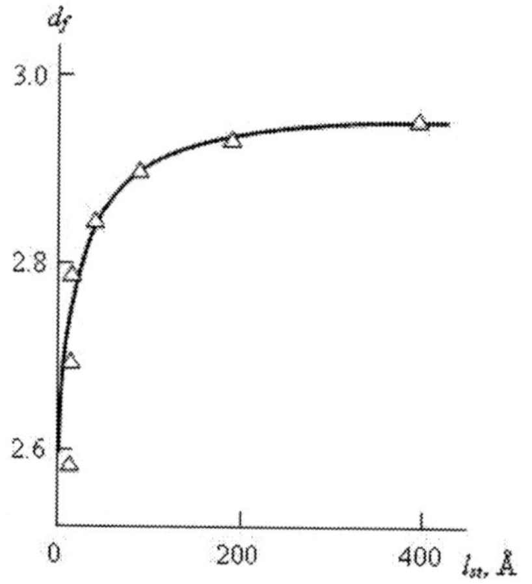

Figure 1.11. The generalized multifractal presentation of carbon plastics structure $d_f(l_{st})$ for l_f=20 mm [32].

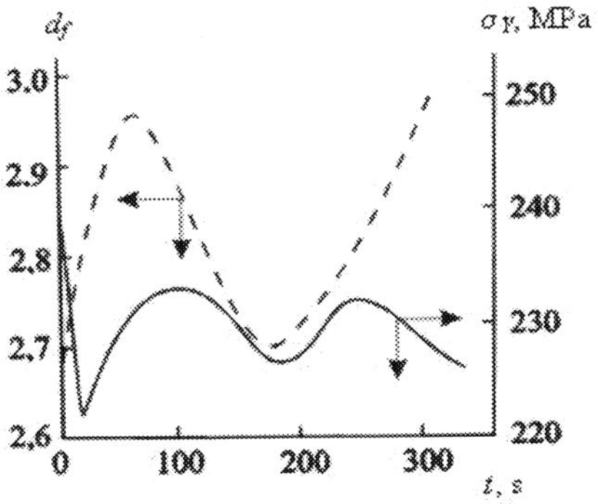

Figure 1.12. The comparison of schematic dependences of structure fractal dimension d_f (1) and yield stress σ_Y (2) on duration t of components blending in rotating electromagnetic field for carbon plastics on the basis of phenylone at l_f=20 mm [32].

The polymers structure formation periodicity was noted repeatedly earlier [33]. However, the methodics of necessary value k (or d_f) goal-directed obtaining was proposed above, which can be realized at a definite combination of t and l_f values. This circumstance allows to obtain carbon plastics with necessary structure and, consequently, with desirable properties complex [32].

Therefore, the observed at carbon plastics preparation by "dry" mode, including components blending in rotating electromagnetic field with the aid of nonequiaxial ferromagnetic particles, sine change of structural characteristic and properties as a function of blending duration is polymers structure synergetic nature manifestation. At present instance structural changes description is fulfilled within the frameworks of a discrete-wave model, although it is possible to use other synergetics approaches. The principal possibility of purposeful regulation of carbon plastics properties is shown as a function of processing in electromagnetic field duration and ferromagnetic particles size.

As it is known [1], at structure self-organization the universal adaptation algorithm corresponds to transition from a structure instability previous point to a subsequent one:

$$A_m = \frac{Z_n}{Z_{n+1}} = \Delta_i^{1/m}, \qquad (1.17)$$

where A_m is structure adaptivity threshold, Z_n and Z_{n+1} are critical values of governing parameter, controlling structure formation, Δ_i is structure stability measure, m is possible reconstructions.

Besides, the equation (1.14) in a general form can be written as follows [7]:

$$\Lambda_n = l_0 C_\infty^n . \qquad (1.18)$$

The physical grounds of large C_∞^{if} increase in comparison with C_∞ for the initial matrix polymer in virtue of the equation (1.18) power character are as follows. CF with relatively smooth surface fix polymer's macromolecules by means of physical and/or chemical interactions, depriving their segments ability to molecular mobility [31]. This is equivalent to

l_{st} growth or C_∞ increase. In such treatment the reconstructions number m' is determined as follows [7]:

$$m' = \frac{C^{if}}{C_\infty} = \frac{C_\infty^n}{C_\infty} = C_\infty^{n-1}. \tag{1.19}$$

As it follows from the data of Figure 1.8, bifurcation point of polymeric matrix structure is reached at $d_f \approx 2.95$, i.e. the greatest d_f value for real physical fractals [10] or at $n=8$ ($d_f = d_{f_8}$). Then in assumption that n^{th} sublevel structure is defined by polymeric matrix initial structure and rotating electromagnetic field action it can be accepted $Z_n = d_{f_n}$, $Z_{n+1} = d_{f_8}$. Thus calculated A_m values and determined according to the table of values Δ_i and m, determining by "gold proportion" law, the indicated parameters for the studied carbon plastics are adduced in table 1.1. In this table the values m', calculated according to the equation (1.19), are also quoted. These results totality allow to make the following conclusions. Firstly, at t increase in the range $0 \le t \le 60$ s structure adaptivity grows up to the greatest value $A_m=1.0$. Secondly, structure stability reduces up to the smallest value $\Delta_i=0.213$. Thirdly, reconstructions number m in structure increases up to $m=128$. As it follows from the data of table 1.1, statistical segments number of initial matrix polymer by statistical segment of nth sublevel or m' increases similarly to m. At small n the condition $m' < m$ is observed, i.e. structure reconstruction possibility is supposed not in the interfacial layer only, but in the bulk polymer matrix and at $n=0$ $m = m' = 128$, i.e. modification of bulk polymeric matrix structure is completely accomplished.

Since in the given context it is supposed that the governing parameter is carbon plastics structure fractal dimension d_f, then the estimation of period doublings number a according to the equation (1.5) is of interest. At $T=293$ K (Figure 1.1) $Z_\infty=2.59$ and $Z_n=2.28$ and then $a=0.76$, and at $T_{av}=400$ K (Figure 1.8) $Z_\infty=2.95$, $Z_n=2.67$ and then $a=0.84$, i.e. again the value a is close to unity. As it follows from Figs. 1.1 and 1.8, the time from $t=0$ up to maximum d_f realization is equal to 60 s and for subsequent reaching of maximum and equal to initial value d_f 120 s requires, i.e. period doubling is observed. Further structure change periodicity violates and system behaviour becomes a chaotic one.

Table 1.1. Synergetic parameters of carbon plastics structure [7]

n	A_m	Δ_i	m	m'
1	0.817	0.380	4	1
2	0.875	0.384	8	2
3	0.915	0.285	16	4
4	0.946	0.255	32	8
5	0.966	0.255	32	16
6	0.983	0.232	64	32
7	0.993	0.213	128	64
8	1.0	0.213	128	128

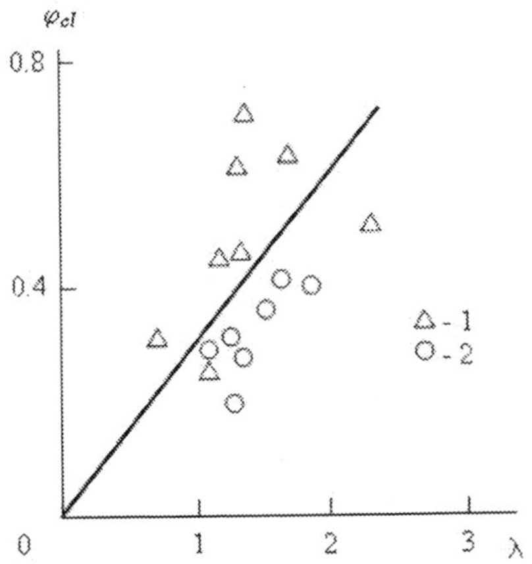

Figure 1.13. The dependence of relative fraction of clusters φ_{cl} on feedback parameter λ for carbon plastics on the basis of phenylone, obtained with magnetic (1) and mechanical (2) separation application [37].

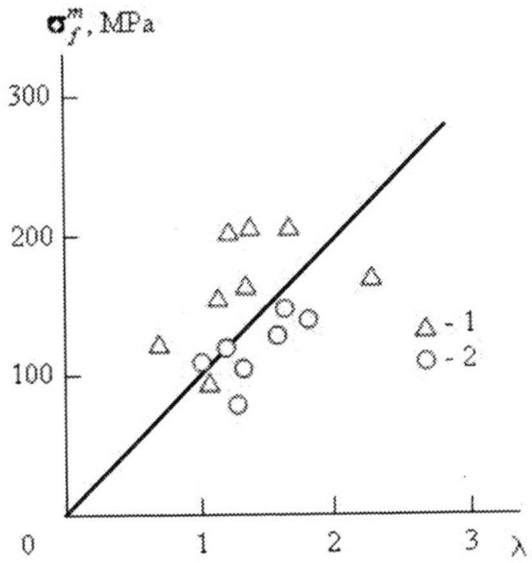

Figure 1.14. The dependence of bulk polymeric matrix strength σ_f^m on feedback parameter λ for carbon plastics on the basis of phenylone. The designations are the same as in Figure 1.13 [37].

It has been shown above, that physical significance of feedback effect according to the equation (1.10) for the considered carbon plastics is limitingly simple: φ_{if} increase results to φ_{cl} reduction and vice versa. The question about technical aspect of feedback arises. Therefore the authors of papers [37, 38] studied the influence degree of feedback in carbon plastics on the basis of phenylone structure on their mechanical characteristics, particularly, on fracture stress (strength).

It was shown above, that a governing parameter at carbon plastics structure formation (more precisely, their interfacial regions) is fibers orientation factor η. Then feedback parameter λ can be calculated according to Puancare equation (1.12), in which indexes n, n+1, ... designate successive intervals of duration t of components blending in rotating electromagnetic field (t_1=5 s, t_2=10 s and so on) [37].

In Figure 1.13 the dependence $\varphi_{cl}(\lambda)$ is adduced, from which φ_{cl} growth at feedback intensification follows, expressed by λ increase. Such picture corresponds completely to the relationship (1.10): λ increase results to polymeric material "pumping" from interfacial regions to bulk polymeric matrix and, as consequence, to local order domains fraction rise in it. In other words, in the very general terms the feedback degree change results to polymeric matrix structure change [38]. It is natural to expect, that the indicated structure change causes polymeric matrix properties variation, specifically, its strength σ_f^m, which can be calculated according to the equation (1.7). As it follows from the data of Figure 1.14, where the dependence $\sigma_f^m(\lambda)$ is adduced, feedback level change tells actually substantially on the bulk polymeric matrix strength: λ growth from about 0.67 up to 2.27, i.e. in three times, results to σ_f^m increase approximately from 78 up to 218 MPa, i.e. in 2.7 times. Let's note the specific feature of linear correlation $\sigma_f^m(\lambda)$: it passes through coordinates origin and this means that in case of feedback absence polymeric matrix will have zero strength.

Proceeding from the stated above observations, one should suppose that occurring in virtue of availability of feedback "pumping" of material from interfacial regions into bulk polymeric matrix changes interfacial regions strength σ_a and also defines interconnection of the stresses σ_f^m and σ_a. σ_a value can be calculated according to the equation (1.7) with replacement of φ_{cl} on φ_{if} and supposing for interfacial regions C_∞=9 according to the indicated above reasons. In Figure 1.15 the relation between strength of bulk polymeric matrix σ_f^m and interfacial regions σ_a is adduced, from which σ_f^m reduction at σ_a growth and vice versa follows. Such interconnection was expected in virtue of effect of feedback and its structural expression: polymer material "pumping" from one densely-packed structural component into another one.

As it is well known [24, 39], interfacial regions properties define to a great extent polymer composites properties as engineering materials. The data of Figure 1.16, in which the dependence of experimentally determined macroscopic strength of carbon plastics σ_f^c on σ_a is adduced, demonstrate clearly this rule: σ_a increase from 35 up to 125 MPa results to σ_f^c growth from 300 up to 406 MPa. Therefore, the combination of Figs. 1.13-1.16 data allows to trace the influence of carbon plastics structural changes, occurring owing to feedback availability, on their mechanical properties: the feedback intensification (λ rise) leads to clusters relative fraction φ_{cl} in bulk polymeric matrix increase (Figure 1.13), its strength σ_f^m growth (Figure 1.14), interfacial regions strength σ_a decrease (Figure 1.15) and in the long run to composite macroscopic strength reduction (Figure 1.16). The practical conclusion from the said above is obvious: for carbon plastics strength rise the feedback parameter must be reduced. So, from the data of Figs. 1.14-1.16 two limiting cases follow: at λ=0 σ_f^m=0, σ_a=180 MPa and σ_f^c=440 MPa and at λ=2.4 σ_f^m=240 MPa, σ_a=28 MPa and σ_f^c=290 MPa, i.e. σ_f^c reduction in about one and a half times at λ increase from 0 up to 2.40 is observed [37, 38].

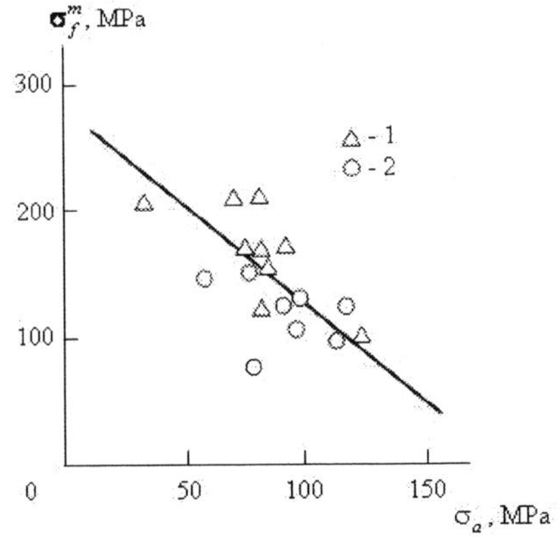

Figure 1.15. The relation between strength of bulk polymeric matrix σ_f^m and interfacial regions σ_a for carbon plastics on the basis of phenylone. The designations are the same as in Figure 1.13 [38].

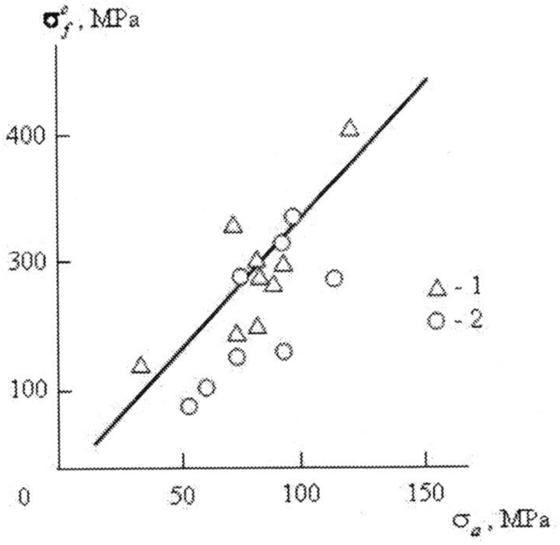

Figure 1.16. The dependence of macroscopic strength σ_f^c on interfacial regions strength σ_a for carbon plastics on the basis of phenylone. The designations are the same as in Figure 1.13 [38].

Taking into account the said above the question arises, how feedback parameter can be purposefully regulated. The answer to this question is given by the graph of Figure 1.17, where the dependence of λ on interfacial regions governing parameter η is adduced. Since some delay (reaction, see the equation (1.12)) of λ in comparison with η can be expected, then in Figure 1.17 this dependence is given as $\lambda_{n+1}(\eta_n)$. From the data of Figure 1.17 it follows, that the value λ can be reduced by fibers orientation factor λ increase. So, for the indicated above σ_f^c increase from 290 up to 440 MPa or λ reduction from 2.40 up to 0 η rise is required from 0.10 up to 0.55 that in common case is attainable on practice result [37].

Therefore, the results obtained above allow to elucidate the structural significance of feedback effect for carbon plastics on the basis of phenylone and to demonstrate its influence on these materials strength. The feedback parameter reduction can result to substantial growth of carbon plastics macroscopic strength. To regulate this parameter value is possible by fibers orientation factor change, which is a governing parameter for interfacial regions [37, 38].

As follows from the adduced above data, polymer composites structure is described by a fractal dimensions number, which is the most general structure informant. However, a polymer represents in itself a specific solid body, containing of long chain macromolecules. This fact defines polymer chain molecular characteristic decisive influence on bulk polymers structure and properties. The equations (1.4), (1.16), (1.18) and so on are given by such influence examples. Proceeding from this, the authors [40] attempted to obtain a single analytical interconnection between structural and molecular characteristics for polymer composites filled with disperse particles and short fibers. As particulate-filled composites two series of composites polyhydroxiether-graphite (PHE-Gr) with nontreated (PHE-Gr-I) and treated by nitric and sulfuric acids in relation 1:1 by volume (PHE-Gr-II) filler surface were selected. The necessary values of fractal dimensions for these composites were accepted according to the data of papers [24, 41]. Besides, carbon plastics on the basis of phenylone were used.

For the studied polymer composites structure description three fractal dimensions were chosen: fractal dimension of filler particles (fibers) network D_n, characterizing their distribution in polymeric matrix [42], fractal dimension of particles (fibers) surface d_{surf}, characterizing its structure (roughly speaking, roughness) [43] and dimension of polymeric matrix excess energy localization energy D_f, characterizing modification ("disturbance") degree of polymeric matrix at filler introduction [41]. As it was noted above, these dimensions for composites PHE-Gr are accepted according to [24, 41] and for carbon plastics they were determined as follows. D_n value was calculated according to the equation [24]:

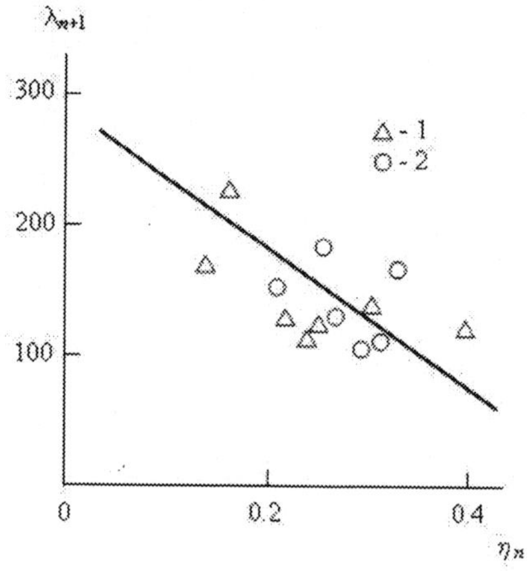

Figure 1.17. The dependence of feedback parameter λ_{n+1} on fibers orientation factor η_n for carbon plastics on the basis of phenylone. The value λ is taken with delay on one temporal interval in comparison with η. The designations are the same as in Figure 1.13 [37].

$$\varphi_{if} = \frac{D_n + 2.55d_0 - 7.10}{4.18}, \qquad (1.20)$$

where d_0 is fractal dimension of particles (fibers) surface of initial (nonaggregated) filler and φ_{if} value was determined according to the equation (1.11).

It was supposed, that for carbon plastics $d_{surf}=d_0=2$ [40]. D_f value can be calculated according to the following equation [10]:

$$D_f = \frac{2(1-\nu)}{1-2\nu}, \quad (1.21)$$

where ν is Poisson's ratio, estimated according to the equation (1.2).

Let's note the important feature of the dimension D_f: it is connected unequivocally with fractal (Hausdorff) dimension of polymeric matrix structure d_f by the relationship [10]:

$$D_f = 1 + \frac{1}{3-d_f}. \quad (1.22)$$

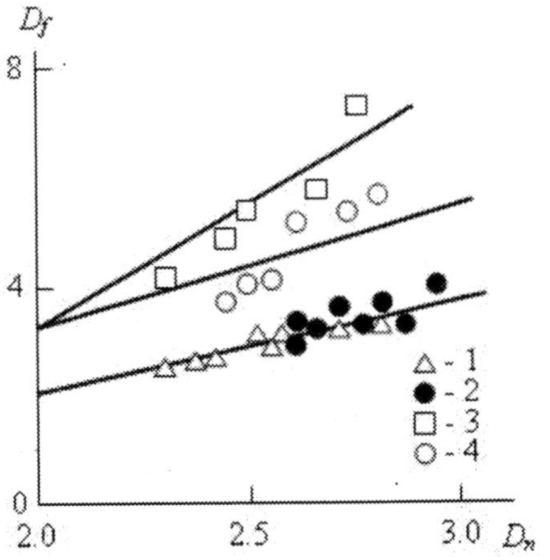

Figure 1.18. The dependences of excess energy localization regions dimension D_f on filler particles network dimension D_n for carbon plastics on the basis of phenylone, obtained with magnetic (1) and mechanical (2) separation application and particulate-filled composites PHE-Gr-I (3) and PHE-Gr-II (4) [40].

In Figure 1.18 the dependences $D_f(D_n)$ for the studied composites are adduced, from which D_f linear growth at D_n increase follows, i.e. at the change of particles (fibers) of filler aggregation and distribution in polymeric matrix. However, for different matrix polymers (PHE and phenylone) these straight lines at D_n=2 are extrapolated to various values D_f^0 (~ 3.4 and 2.2, respectively). Since D_f increase means intensification of polymeric matrix structure "disturbance", the structural significance of which is the decrease of relative fraction of local order domains (clusters) φ_{cl} [41], then D_f^0 value corresponds to the greatest possible value φ_{cl}^{max}, which can be determined according to the equation (1.3) at $\beta_T=\beta_p=0.4$ and then minimum dimension d_f^{min} can be calculated according to the equation (1.4) at $\varphi_{cl}=\varphi_{cl}^{max}$. With such methodics using d_f^{min}=2.597 for PHE and d_f^{min}=2.294 for phenylone were obtained, that according to the equation (1.22) gives D_f^0=3.48 and 2.42 for these polymers. It is easy to see, that the calculated values D_f^0 correspond well enough to extrapolation of Figure 1.18. Therefore, the equations (1.3) and (1.4) define interconnection of matrix polymer molecular characteristics (at the indicated above conditions β_T=β_p=0.4 and φ_{cl}=φ_{cl}^{max}) with composite structural characteristics.

Besides, from the data of Figure 1.18 it follows, that composites PHE-Gr-I and PHE-Gr-II at the same value D_f^0 have different slopes of the linear dependences $D_f(D_n)$, that can be attributed to higher d_{surf} values for the first of them [43]. For this supposition confirmation on Figure 1.19 the dependence of value ΔD_f=D_f- D_f^0 on Δd_{surf}=d_{surf}-d_0 is shown, where for graphite d_0=2.17 [44]. As one can see, for composites PHE-Gr this dependence is approximated by one linear correlation that confirms the made supposition. Analytically this correlation can be expressed as follows [40]:

$$D_f - D_f^0 = 5.2\left(d_{surf} - d_0\right). \qquad (1.23)$$

Since for carbon plastics d_{surf}=d_0=2 was accepted, then this supposes that D_f change for them is due to an other reason. If to take into accout, that as filler for phenylone short fibers are used, the orientation of which changes in virtue of the applied blending technology specific features, then it can be supposed that the indicated reason will be fibers orientation factor η, which is the governing parameter for carbon plastics structure [45]. In Figure 1.19 the dependence $\Delta D_f(\Delta\eta)$ for carbon plastics is also adduced, where minimum value η=0.142 [45]. It is significant that this dependence is approximated by linear correlation also, coinciding with the correlation $\Delta D_f(\Delta d_{surf})$ and is analytically described as follows [40]:

$$D_f - D_f^0 = 5.2(\eta - 0.142). \qquad (1.24)$$

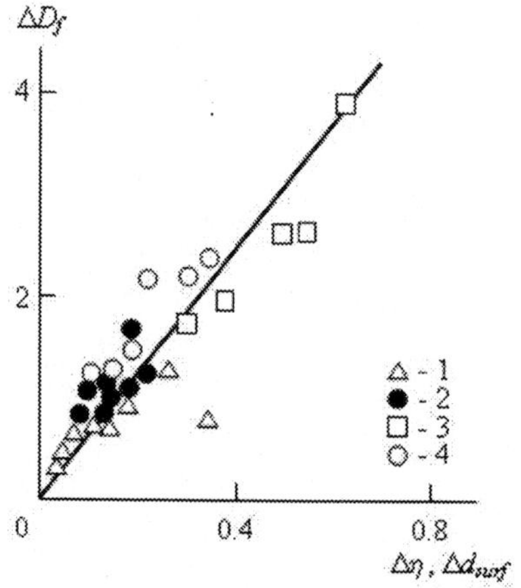

Figure 1.19. The dependence of excess energy localization regions dimension ΔD_f change on the change of fibers orientation factor $\Delta\eta$ (1, 2) and filler particles surface fractal dimension Δd_{surf} (3, 4). The designations are the same as in Figure 1.18 [40].

Therefore, d_{surf} change for particulate-filled composites and η change for composites, filled with short fibers, influence equally on the polymeric matrix "disturbance" level, characterizing by dimension D_f variation [40].

The stated above results allow to obtain generalized interconnection of fractal dimensions D_f (or d_f), d_{surf} and D_n, accounting for matrix polymer molecular characteristics influence by the parameter D_f^0 introduction. In Figure 1.20 the dependences D_f/d_{surf} on D_n for the studied composites are adduced, which can be approximated by the following generalized equation [40]:

$$\frac{D_f}{d_{surf}} = \frac{D_f^0}{d_0} + 1.41\times10^{-3} D_n^6, \quad (1.25)$$

where d_0 value in all cases is accepted equal to 2.

The equation (1.25) allows to estimate polymeric matrix structure "disturbance" degree, characterized by the value D_f or d_f, if the values of glass transition temperature T_g and matrix polymer molecular characteristics S and C_∞ as well, and dimensions d_{surf} and D_n, characterizing the structure of filler in polymer composite. Let's note, that parameters T_g, S and C_∞ are connected between each other by the empirical relationship [4]:

$$T_g \approx 191\left(\frac{S}{C_\infty}\right)^{1/2}, \text{ K.} \quad (1.26)$$

d_f experimental values can be determined according to the equation (1.1) and theoretical d_f^T - according to the equations (1.25) and (1.22). In Figure 1.21 the comparison of values d_f and d_f^T for the studied composites is adduced, from which their excellent correspondence follows – the average discrepancy of theory and experiment by fractal dimension fractional part is less than 2 %.

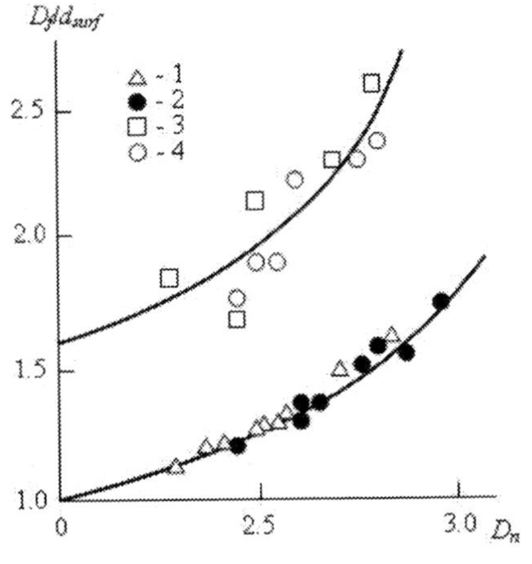

Figure 1.20. The dependences of dimensions ratio D_f/d_{surf} on filler particles network dimension D_n. The designations are the same as in Figure 1.18 [40].

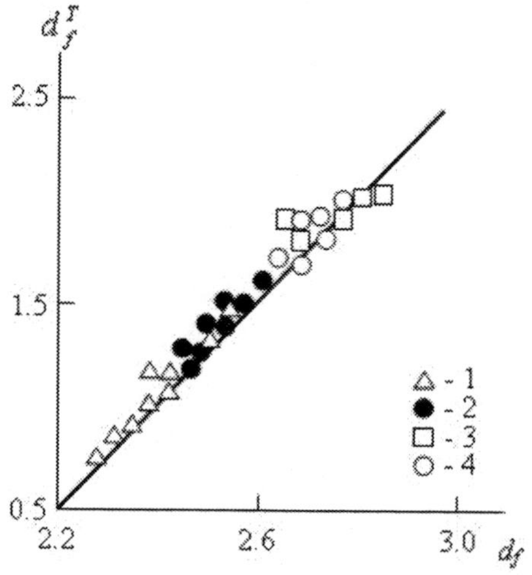

Figure 1.21. The comparison of theoretical d_f^T and experimental d_f values of polymeric matrix structure fractal dimension. (The explanations are in the text). The designations are the same as in Figure 1.18 [40].

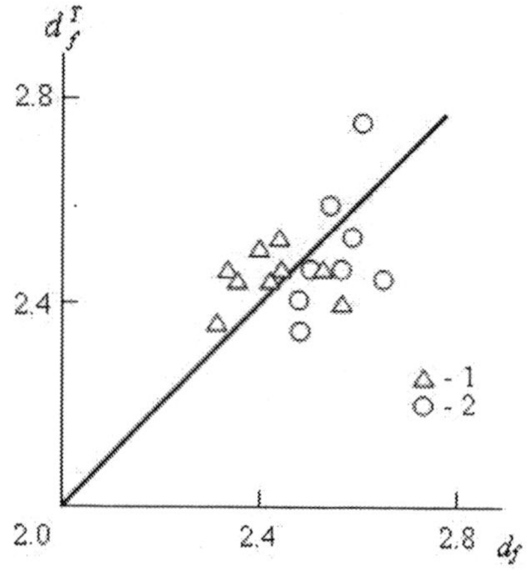

Figure 1.22. The comparison of structure fractal dimensions d_f^T (the equation (1.4)) and d_f (the equation (1.1)) for carbon plastics on the basis of phenylone, obtained with magnetic (1) and mechanical (2) separation application. The straight line shows relation 1:1 [46].

Therefore, the stated above results demonstrated the community of the change of structural characteristics of composites, filled with disperse particles and short fibers. The complexity of these materials structure requires for their description, as minimum, four independent order parameters, from which two are related to polymeric matrix and two – to filler. The fractal analysis methods application allows to obtain the relationship, describing these parameters interconnection. Specific features of either filler type are reflected in the choice of one order parameter for them: in case of disperse particles this is fractal dimension of filler particles surface, characterizing their aggregation degree, and in case of short fibers this is their orientation factor [40].

And in conclusion of the section let's consider briefly estimation methodics of composites structure fractal dimension d_f, using the stated above synergetics principles, in particular, the equation (1.10) [46]. Hawing calculated interfacial regions relative fraction φ_{if} according to the equation (1.11), one can estimate φ_{cl} value from the equation (1.10) and then to calculate theoretical value d_f (d_f^T) according to the equation (1.4). And as earlier, the experimental values d_f were calculated according to the formula (1.1). In Figure 1.22 the comparison of the values d_f^T and d_f, calculated by the indicated model, for two series of carbon plastics on the basis of phenylone is adduced. As one can see, the good enough correspondence of these parameters is obtained (the average discrepancy by fractal dimension fractional part is equal to ~ 18 %). If to take into consideration, that calculation d_f precision by mechanical parameters does not exceed 10 % then both indicated methods of d_f determination can be regarded as equal [46].

The opposite calculation variant is quite possible, i.e. the value φ_{if} estimation by the known values d_f, C_∞ and S. Such calculation is fulfilled in the following succession: according to the equation (1.2) Poisson's ratio is calculated, then according to the equation (1.1) dimension d_f hen according to the equation (1.10), further according to the formula (1.4) φ_{cl} value and at last – the value φ_{if}^T the comparison of the values calculated by the pointed methods is adduced in Figure 1.23. In this case the values φ_{if} and φ_{if}^T discrepancy is larger (~ 34 %), but the obtained tendency of these parameters change and symmetrical arrangement of the data points relatively to a straight line, giving the relation 1:1, suppose that the error increase is due to inaccuracy of the data usage in the equations (1.1), (1.2), (1.4) and (1.10) [46].

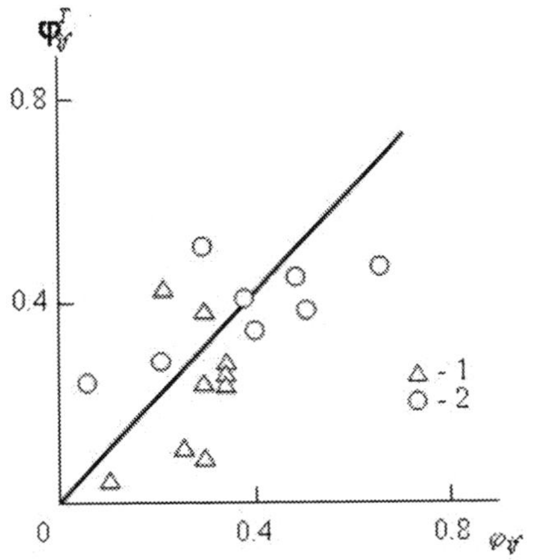

Figure 1.23. The comparison of interfacial regions relative fraction values φ_{if}^T (the equation (1.10)) and φ_{if} (the equation (1.11)) for carbon plastics on the basis of phenylone, obtained with magnetic (1) and mechanical (2) separation application. The straight line shows relation 1:1 [46].

1.2. Formation Mechanisms and Structure of Interfacial Regions

Before to proceed to the main theme of the present section, let's consider methodics of interfacial regions two main structural characteristics estimation, namely, their relative fraction in composite structure and fractal dimension, without determination of which quantitative description of the indicated regions is impossible.

In paper [24] on the example of two series of particulate-filled composites PHE-Gr it has been shown, that interfacial regions relative fraction φ_{if} is defined by two factors: fractal dimensions of particles (aggregates of particles) surface d_{surf} and filler particles network D_n, characterizing degree of polymeric matrix filling by filler [42]. The indicated interconnection is described analytically by the empirical equation (1.20). This equation describes well φ_{if} change as a D_n function for composites PHE-Gr, but the attempt to apply it for the value φ_{if} estimation in case of carbon plastics on the basis of phenylone gives a large error. The indicated discrepancy is explained by the fact that at the equation (1.20) derivation the following relationship was used [24]:

$$\varphi_{if} = 0.61(d_{surf} - 2), \qquad (1.27)$$

which is correct for composites PHE-Gr, but is not correct for carbon plastics, which are not supposed to have appreciable aggregation of filler particles (carbon fibers) and consequently $d_{surf}=d_0=$const, although φ_{if} value varies in about 6 times. Therefore the authors [47] proposed generalized methodics of interfacial regions contents calculation, working equally well for both particulate-filled composites and for composites filled with short fibers.

The value of filler particles network fractal dimension D_n for carbon plastics on the basis of phenylone can be estimated as follows. As it was shown in papers [48-50], for polymer composites the following condition should be fulfilled:

$$\nu_F = \frac{2.8}{2 + D_n}, \qquad (1.28)$$

where ν_F is Flory exponent, connected with fractal dimension of macromolecular coil in diluted solution D_c by a simple relationship [51]:

$$\nu_F = D_c^{-1}. \qquad (1.29)$$

In its turn, the values D_c and fractal dimension d_f of polymers condensed state are connected with each other by the equation [52]:

$$d_f = 1.5 D_c. \qquad (1.30)$$

And as before, the values d_f and φ_{if} were estimated according to the equations (1.1) and (1.11), accordingly.

In Figure 1.24 the dependence $\varphi_{if}(D_n)$ for carbon plastics on the basis of phenylone is shown and analogous dependence for composites PHE-Gr according to the data of paper [24] is also adduced. As one can see, by the reason indicated above these dependences differe, although they are linear ones. To take into account the influence of filler particles surface structure, characterized by its fractal dimension d_{surf}, the authors [47] plotted the dependence of the generalized variable $d_{surf}\varphi_{if}$ on D_n, which is shown in Figure 1.25. Now the data for both composite types, filled with short fibers and disperse particulates, are set on the single straight line, passing through coordinates origin. The correlation shown in Figure 1.25 is expressed analytically as follows [47]:

$$\varphi_{if} = \frac{1.20(D_n - 2)}{d_{surf}}. \tag{1.31}$$

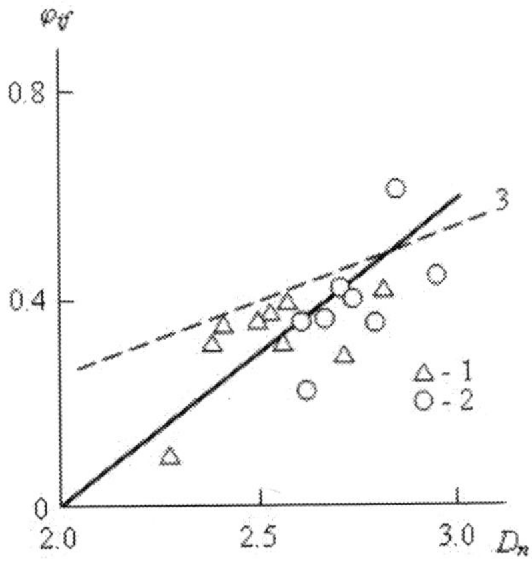

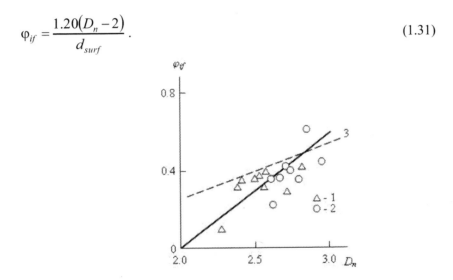

Figure 1.24. The dependences of interfacial regions relative fraction φ_{if} on fractal dimension of filler particles network D_n for carbon plastics on the basis of phenylone, obtained with magnetic (1) and mechanical (2) separation application and also for composites PHE-Gr according to the data of paper [24] (3) [47].

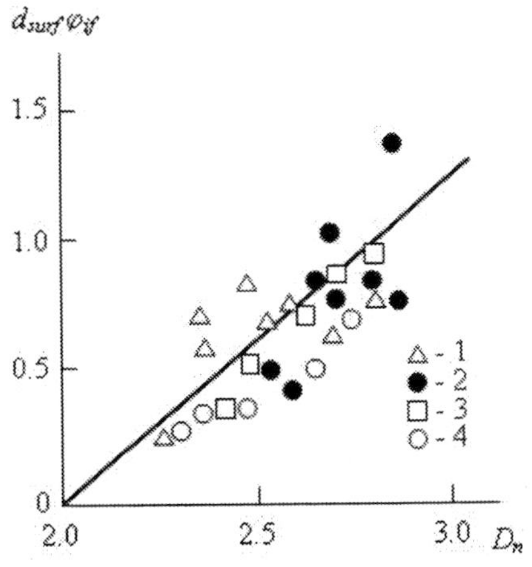

Figure 1.25. The dependence of the generalized variable $d_{surf}\varphi_{if}$ on fractal dimension of filler particles network D_n for carbon plastics on the basis of phenylone, obtained with magnetic (1) and mechanical (2) separation application and also for composites PHE-Gr-I (3) and PHE-Gr-II (4) [47].

Hence, φ_{if} value in polymer composites filled with disperse particulates and short fibers, is defined by two factors: filler particles (fibers) distribution character in polymeric matrix, described by the dimension D_n, and filler particles (aggregates of particles) surface structure, characterized by the dimension d_{surf}. The equation (1.31) can be used for φ_{if} value calculation, since dimensions D_n [42] and d_{surf} [53] can be determined experimentally. In Figure 1.26 the comparison of the obtained experimentally φ_{if} and calculated according to the equation (1.31) φ_{if} values of interfacial regions relative fraction for carbon plastics on the basis of phenylone and composites PHE-Gr is adduced. As one can see, between the experiment and theory good enough correspondence is obtained (the average discrepancy φ_{if} and φ_{if}^{T} makes less than 20 %, that is considerably better than the obtained one in paper [46]).

As it was shown above, the structure of carbon plastics on the basis of phenylone was a synergetic system, for which carbon fibers orientation factor η serves as a governing parameter [45]. η increase results to d_f growth and according to the equations (1.28)-(1.30) – to D_n rise. The following relationship was obtained between parameters η and D_n [47]:

$$\eta = 0.506(D_n - 2), \tag{1.32}$$

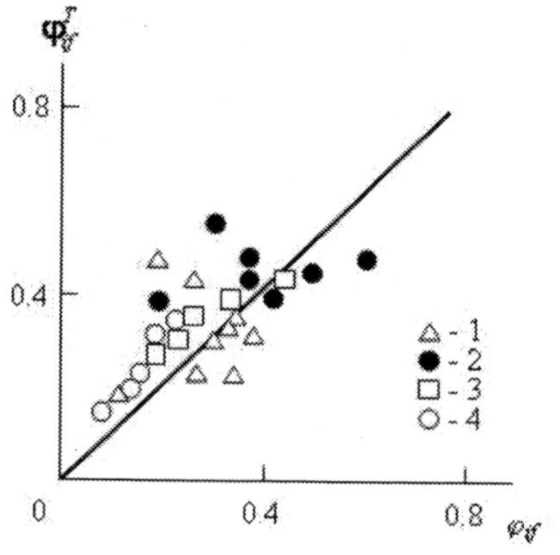

Figure 1.26. The comparison of the determined experimentally φ_{if} and calculated according to the equation (1.31) φ_{if}^{T} interfacial regions relative fraction. The designations are the same as in Figure 1.25 [47].

which, after its substitution in the equation (1.31), allows to obtain the dependence of φ_{if} on η in form [47]:

$$\varphi_{if} = 1.09\eta . \tag{1.33}$$

In Figure 1.27 the comparison of φ_{if} and estimated according to the equation (1.33) φ_{if}^{T} values of interfacial regions relative fraction for carbon plastics is adduced on the basis of phenylone. As one can see, the good enough correspondence of theory and experiment is obtained again. Such correspondence was expected, since, strictly speaking, fibers orientation factor was a governing parameter precisely for interfacial regions [45].

Therefore, the results stated above have shown that interfacial regions relative fraction in polymer composites, filled with disperse particulates or short fibers, is defined by two structural factors: filler distribution in polymeric matrix and its surface structure. Both indicated factors can be characterized with the aid of corresponding fractal dimensions. The empirical equations are obtained, allowing the calculation of interfacial regions relative fraction as a function of the indicated dimensions and for carbon plastics – as a function of governing parameter.

As it was noted above, interfacial layers structure and properties define polymer composites macroscopic properties in many respects. The importance of problem defined the application for interfacial layers study such modern physical conceptions as fractal analysis, irreversible aggregation models and so on [24, 39]. The modern studies in the field of surface chemistry and physics allow to describe processes, connected with surface energy role in structure and properties formation for nanometer scale substance. The universal informant of the substance structural state in both animate and inanimate nature is the fractal dimension [54]. A bulk topological dimension d is equal to 3 and the surface to 2. The transition from bulk to surface is characterized by the interfacial layer fractal dimension d_f^{if}. In paper [54] the surface transitions from d=3 to d=2 major properties concept is developed, in correspondence to which at the reduction of dimension value of the substance filling by three-dimensional space at transition from bulk part of material object on its surface the energy is released, which is found experimentally as the condensed phase surface energy. In addition the surface energy value is defined by the difference d- d_f^{if}, where d is topological dimension of space, which is equal to 3 and d_f^{if} changes within the limits $2\leq d_f^{if}<3$ [54]. It is obvious, that for quantitative description of interfacial layers properties within the frameworks of modern notions the value d_f^{if} determination methodics is necessary. The authors [55] elaborated two variants of such methodics and compared the obtained results with their aid on the example of carbon plastics on the basis of phenylone.

As Lipatov has shown [30], the special plase in polymer composites study calorimetric methods take up. According to Wunderlich rule the heat capacity increment at glass transition $\Delta C_p^{'}$ in counting on one mole of macromolecule minimum (by sizes) structural elements is a constant value, which is equal to 12.2-17.3 kJ/(moles·K), and is calculated according to the formula [30]:

$$\Delta C_p^{'} = \overline{M}\Delta C_p^c, \qquad (1.34)$$

where $\overline{M}$ is average molecular weight of the element, $\Delta C_p^{'}$ is an experimentally observed jump of heat capacity at constant pressure at glass transition temperature.

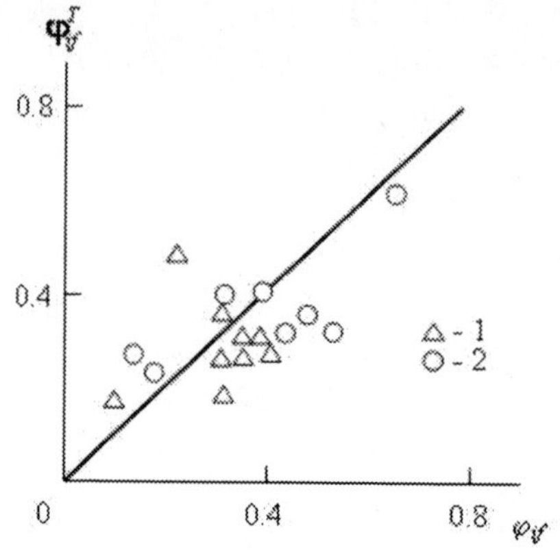

Figure 1.27. The comparison of the determined experimentally φ_{if} and calculated according to the equation (1.33) φ_{if}^T interfacial regions relative fraction for carbon plastics on the basis of phenylone, obtained with magnetic (1) and mechanical (2) separation application [47].

The indicated minimum structural element of macromolecule is a statistical segment of length l_{st}, determined according to the equation (1.14). Simulation of the statistical segment by cylinder allows to determine its volume V_{st} as follows [55]:

$$V_{st} = l_{st}S = l_0 C_\infty S . \qquad (1.35)$$

And at last, the relation between statistical segment volume V_{st} and its molecular weight $\overline{M}$ is given by the equation [11]:

$$V_{st} = \frac{\overline{M}}{\rho N_A}, \qquad (1.36)$$

where ρ is material density, N_A is Avogadro number.

The combination of the equations (1.14) and (1.34)-(1.36) allows to obtain the following relationship for the value C_∞ of interfacial layers estimation [55]:

$$C_\infty = \frac{\Delta C_p^{'}}{\Delta C_p^c \rho N_A l_0 S m_m}, \qquad (1.37)$$

where m_m is phenylone mole mass, which is equal to 0.243 kg/kmole [56].

The equation (1.37) with reference to phenylone at using of the values $\Delta C_p^{'}$=12.2 kJ/(mole·K) and ρ=1400 kg/m^3 and the indicated above magnitudes l_0 and S also comes to the following simple equation [55]:

$$C_\infty = \frac{2.7}{\Delta C_p^c}, \qquad (1.38)$$

where ΔC_p^c is given in kJ/(kg·K).

Further the fractal dimension d_f^{if} estimation can be fulfilled with the aid of the relationship (1.16) at the corresponding replacement of d_f on d_f^{if}. Hereinafter determined according to the equation (1.16) d_f^{if} value will be designated as $d_{f\,1}^{if}$.

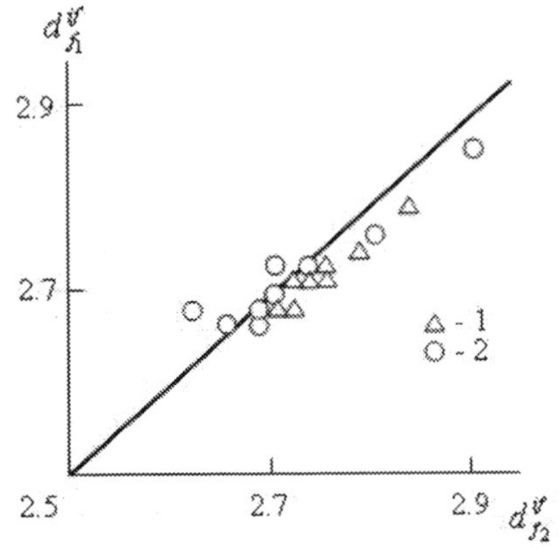

Figure 1.28. The comparison of interfacial layers structure fractal dimensions, calculated according to the equations (1.14) ($d_{f\,1}^{if}$) and (1.4) ($d_{f\,2}^{if}$), for carbon plastics on the basis of phenylone, obtained with magnetic (1) and mechanical (2) separation application. The straight line shows the relation 1:1 [55].

The second method of calculation d_f^{if} ($d_{f\,2}^{if}$) consists in the following. As it is known [31], polymer chains stretch on filler fibers smooth surface, forming densely-packed interfacial regions. This circumstance allows to use for $d_{f\,2}^{if}$ calculation the equation (1.4) with φ_{cl} replacement on φ_{il}, the value of which was calculated according to the equation (1.11). Let's note, that the equation (1.4) application for d_f^{if} calculation means supposition of denser packing of interfacial layers in comparison with bulk polymeric matrix [55].

In Figure 1.28 the comparison of the values d_f^{if} ($d_{f\,1}^{if}$ and $d_{f\,2}^{if}$), calculated according to the described above methodics, is adduced. As one can see between these values of interfacial layers structure fractal dimension the good correspondence is observed (the average discrepancy over fractional part of d_f^{if}, which is an informant of polymeric substance state in interfacial layers [54], makes up ~ 3.5 %).

Therefore, two variants of interfacial layers structure fractal dimension calculation for polymer composites on the basis of the experimental calorimetric data were considered above. Both variants have given concerted results, that supposes their correctness. Let's note, that second methodics correctness, based on the using of the equation (1.4), confirms the supposition of a denser packing of interfacial layers in comparison with bulk polymeric matrix [55].

The development during last decades of such modern physical conceptions as synergetics and fractal analysis allows to obtain principally new notions about formation mechanisms and structure of interfacial regions in polymer composites. It was found out [1] that interfacial regions were system (composite structure) adaptator to external influence (for example, deformation). An interfacial regions adaptive properties nature is due to self-organization of transitional layer on separation boundary filler-polymeric matrix. The functional role of interfacial regions in composites is connected with ensuring of interconnection of bulk polymeric matrix and filler surface, having topological dimensions of 3 and 2, respectively.

It was shown [1, 54], that at the reduction of fractal dimension of substance distribution in space, i.e. at the transition from bulk matrix to surface, an energetic constituent of system increases at the expence of the rise of matter distribution difference value ($3-d_{surf}$) over transitional surface layer thickness. The inalienable characteristic of interfacial regions state is properties gradient, which serves as an informant about layer structure. Proceeding from this, the authors [57] considered polymer composites interfacial regions structure and properties within the frameworks of synergetics and fractal analysis on the example of carbon plastics on the basis of phenylone.

As it was noted above, polymer composites structure real (fractal) dimensions departure from the corresponding topological values results to system energetic constituent change [1]. Therefore as such departure measure the authors [57] used the difference of fractal dimensions of bulk polymeric matrix d_f and interfacial regions d_f^{if} [57]:

$$\Delta d_f = d_f^{if} - d_f, \tag{1.39}$$

where d_f^{if} value was determined with the aid of the described above methodics according to the equation (1.4) [55] and d_f was calculated according to the equation (1.1). And at last, interfacial layer thickness l_{if} can be estimated, using the relationship [58]:

$$\left(\frac{l_{if} + r_{fib}}{r_{fib}} - 1\right)^3 = \varphi_{if}\left(\frac{\varphi_f}{1-\varphi_f}\right), \tag{1.40}$$

where r_{fib} is a filler particle (fiber) radius, φ_f is filler volume contents.

In Figure 1.29 the dependence $l_{if}(\Delta d_f)$ for both series of carbon plastics is adduced, from which l_{if} reduction at Δd_f growth follows. At Δd_f=0, i.e. at condition $d_f^{if}=d_f$, l_{if} value is the greatest (from the physical point of view this means that bulk polymeric matrix and interfacial regions are structurally indistinguishable), and at Δd_f=0, i.e. in case of corresponding fractal and topological dimensions equality (d_f^{if} =2 and d_f=3) interfacial regions are not formed at all. In practice this means, that the condition l_{if}=0 can be realized for ideal rubber and absolutely (microscopically) filler smooth surface. Since in reality the first condition, as minimum, is not realized (for real solids the greatest value d_f=2.95 [10]), then this means obligatory formation of interfacial layers in polymer composites. The similar conclusion was obtained in the particulate-filled composites case [24].

Gibbs was considered interfacial surface as finite by thickness layer, where constitution and thermodynamic properties were different from the same for bulk bordering phases. This allows to express interfacial regions substance thermodynamic state features with the aid of free energy value or, more precisely, of its change ΔG^{if}. ΔG^{if} value in relative units can be obtained from the equation [59]:

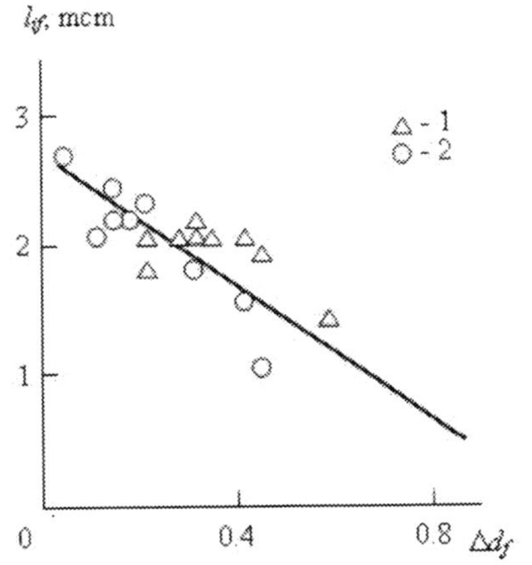

Figure 1.29. The dependence of interfacial layer thickness l_{if} on interfacial regions and polymeric matrix fractal dimensions difference Δd_f for carbon plastics on the basis of phenylone, obtained with magnetic (1) and mechanical (2) separation application [57].

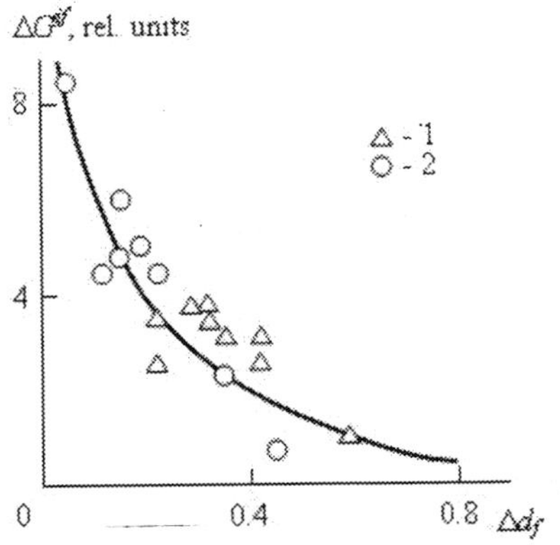

Figure 1.30. The dependence of interfacial regions free energy change ΔG^{if} on interfacial regions and polymeric matrix fractal dimensions difference Δd_f for carbon plastics on the basis of phenylone. The designations are the same as in Figure 1.29 [57].

$$\Delta G^{if} \sim SC_{\infty}\left(3-d_f^{if}\right)^2. \tag{1.41}$$

In Figure 1.30 the dependence $\Delta G^{if}(\Delta d_f)$ is shown, from which the fast decay ΔG^{if} at Δd_f growth follows. This corresponds completely to the stated above postulate about system energetic constituent increase at substance distribution change over interfacial regions thickness [1]. The interfacial regions structure energetic constituent is closely connected with their practically important characteristics – strength σ_a, that follows from the well-known Leidner-Woodhams equation [60]:

$$\sigma_f^c = (\sigma_a + 0.83\tau_m)\varphi_f + \sigma_a K_s(1 - \varphi_f), \quad (1.42)$$

where τ_m is shearing strength of polymeric matrix, K_s is stress concentration coefficient and σ_a value can be calculated according to the equation (1.7).

As it follows from the graph of Figure 1.31, Δd_f increase results to sharp reduction of interfacial layer strength and at Δd_f=1.0 σ_a=0. In other words, in Euclidean objects case, i.e. d_f^{if} =2 and d_f=3, composite strength will be defined according to the equation (1.42) by shearing strength of polymeric matrix only. For the considered carbon plastics in this case σ_f^c value should be equal to about 110 MPa, whereas the experimental σ_f^c values are much higher and are varied within the limits of 195-269 MPa for the case of tension testing. This simple example clearly demonstrates both interfacial regions and structural components fractal nature role for polymer composites in their macroscopic properties formation. Let's note the dependences $\Delta G^{if}(\Delta d_f)$ and $\sigma_a(\Delta d_f)$ similarity, which is obvious for comparison of the graphs of Figs. 1.30 and 1.31.

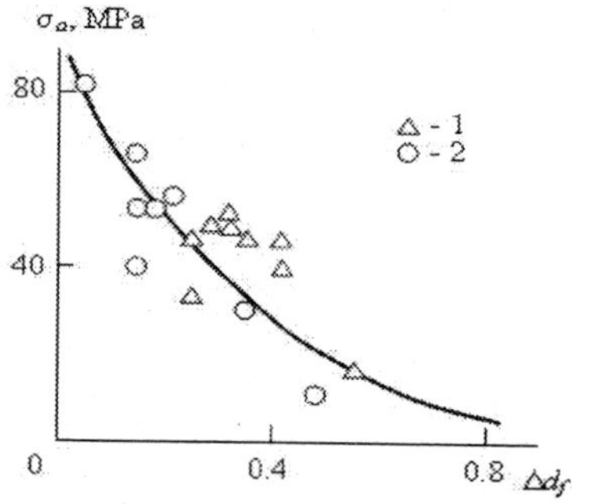

Figure 1.31. The dependence of interfacial layer strength σ_a on interfacial regions and polymeric matrix fractal dimensions difference Δd_f for carbon plastics on the basis of phenylone. The designations are the same as in Figure 1.29 [57].

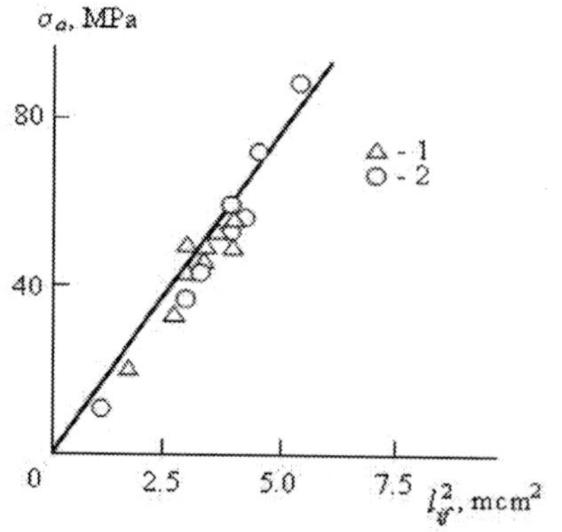

Figure 1.32. The dependence of interfacial layer strength σ_a on its thickness l_{if} for carbon plastics on the basis of phenylone. The designations are the same as in Figure 1.29 [57].

Let's consider further interfacial regions thickness l_{if} and strength σ_a interconnection. In Figure 1.32 the dependence $\sigma_a(l_{if}^2)$ is shown, which is turns out to be linear and passing through coordinates origin. In other words, for the considered composites l_{if} increase raises interfacial regions strength fast, that contributes to composites macroscopic strength growth according to the equation (1.42). It is interesting to compare these data with the obtained earlier similar results. For particulate-filled composites PHE-Gr σ_a reduction at l_{if} growth was found out [26] and for polymers pair (polystyrene and poly-p-methylstyrene) interfacial layer strength linear growth at l_{if}^2 increase was observed [61]. This discrepancy can be explained, as minimum, by two reasons. Firstly, by interfacial regions formation mechanisms difference. For polymer-polymer pair [61] and the studied carbon plastics interfacial regions formation is realized mainly by diffusive processes (see below) and for particulate-filled composites PHE-Gr such processes role is minimum [62]. Secondly, for the considered carbon plastics the fibers relative smooth surface ($d_{surf}\to 2$) contributes to macromolecules stretching on it and to formation densely-packed interfacial regions [31]. For composites PHE-Gr d_{surf} high values (>2.5) result to macromolecular coil conformations conservation, existing in melt [31], that creates conditions for interfacial regions structure loosening [63]. In other words, for carbon plastics l_{if} increase contributes to the indicated regions structure densification, whereas for particulate-filled composites PHE-Gr l_{if} rise gives the opposite effect. In the long run the indicated factors define different type of dependences $\sigma_a(l_{if})$ for carbon plastics and polymer-polymer pair on the one hand and particulate-filled composites PHE-Gr on the other hand.

Table 1.2. The quantitative characteristics of structure of glassy plastics on the basis of polyarylate filled with glassy fiber of mark M [64]

The glassy fibers volume contents	Fiber sizes, mcm	
	diameter	length
0.03	12-20	120-500
0.09	12-20	110-400
0.13	12-18	100-200
0.17	12-15	80-210

Therefore, the stated above results demonstrated the synergetic general principles applicability for interfacial regions structure and properties in polymer composites. The limiting characteristics calculation allows to show clearly again the importance of these regions and composites structural components fractal nature in their macroscopic properties formation. Putting these rules into practice allows to regulate purposefully such important property of composites as their strength [57].

The authors [64] studied composites on the basis of polyarylate DV-102, filled with glassy fiber of alkaliless composition (high-modulus beryllium-calcium-silicate fiber of mark M), the properties of which were adduced in paper [58]. Since glassy fiber is very sensitive to mechanical influence, then in the indicated glassy plastics preparation process high-modulus fiber M is fractured by brittle mode splitting on smaller fragments, as a result of this apart from glassy fiber dispersion its diameter reduction up to minimum value 12 mcm occurs (table 1.2).

Successful usage of fractal analysis methods for interfacial layers structure and properties description in particulate-filled polymer composites [24, 39] allows to expect the indicated approach applicability in case of polymers, filled with short fibers. In paper [58] interfacial layer relative fraction φ_{if} and thickness l_{if} increase in glassy plastics on the basis of polyarylate at glassy fiber contents φ_f rise was found out. Within the frameworks of continuous approach l_{if} value can be calculated according to the equation [65]:

$$l_{if} = 2R_p\left[\left(\frac{\eta_{lp}}{\varphi_f}\right)^{1/3} - 1\right], \tag{1.43}$$

where R_p is a filler particle radius, η_{lp} – lattice packing density, which is accepted equal to ~ 0.74 [65].

In the glassy plastics case filler particles (aggregates of particles) R_p for particulate-filled composites should be replaced by characteristic size of filler particle (fiber) L_{ch}. As L_{ch} for glassy fiber can be accepted either its diameter D_{gf} or its length l_{gf}. As it follows from the data of table 1.2, these both parameters are decreased at φ_f growth. Therefore, from the equation (1.43) l_{if} decrease (and, hence, φ_{if} decrease) at φ_f growth follows at any choice of L_{ch}: $L_{ch}=D_{gf}$ or $L_{ch}=l_{gf}$. In other words, as in case of particulate-filled composites [66] continuous approach is not capable to describe the observed experimentally function $\varphi_{if}(\varphi_f)$ change for glassy plastics. Therefore for the indicated purpose the fractal approach was used, proposed in paper [67], where Yakubov considered sorptional phenomena thermodynamic on fractal objects and proposed the following relationship for the determination of adsorbed phase relative fraction on such object, which in the first approximation was necessary to consider as interfacial regions relative fraction φ_{if}:

$$\varphi_{if} = \rho_p^{-1} m_0^{(1-3/D)}, \tag{1.44}$$

where ρ_p is separate particle density, m_0 is its mass, D is object fractal dimension.

In paper [67] it is implied that as D one must accept fractal dimension of filler particles aggregates. However, as Pfeifer showed [31, 68], decisive influence on the adsorbed layer formation did not have adsorbing object fractal dimension, but its surface dimension d_{surf}. Therefore in the subsequent estimations $D=d_{surf}$ will be accepted [64]. In paper [58] φ_{if} value was calculated according to the equation (1.11), m_0 values can be estimated according to the data of table 1.2 and at the condition ρ_p=const plotting of the dependence of φ_{if} (table 1.3) on glassy fiber volume V_{gf} in double logarithmic coordinates, corresponding to the equation (1.44), in case of its linearity allows to determine the exponent in the indicated relationship from linear graph slope and, hence, the dimension d_{surf}. In Figure 1.33 such graph for the considered glassy plastics is adduced, which actually is linear. From the data of this graph the value $D=d_{surf}=2.14$ was determined. Thus determined d_{surf} value excellently corresponds to the independent experimental estimation $d_{surf}=(2.02\text{-}2.15)\pm0.06$ for various kinds of glass and quartz [44]. Besides, the condition d_{surf}=const at φ_f variation supposes glassy aggregation absence [24], that glassy plastics microphotographs also confirm [25].

Table 1.3. The characteristics of interfacial layer of glassy plastics on the basis of polyarylate [58]

The glassy fibers volume contents	φ_{if}	l_{if}, nm
0.03	0.51	2.5
0.09	0.61	10.2
0.13	0.73	17.4
0.17	0.80	37.8

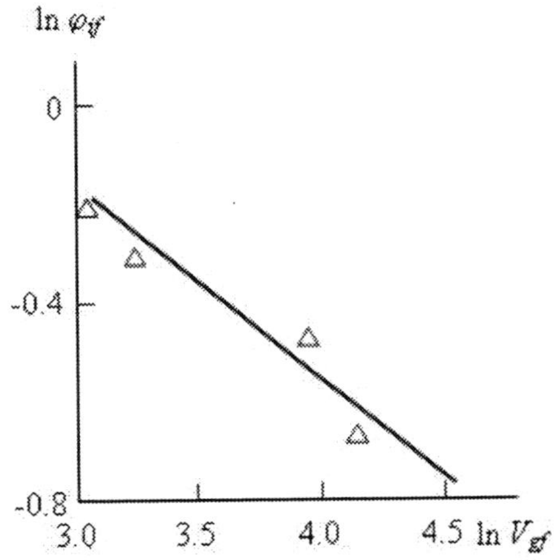

Figure 1.33. The dependence of interfacial layer relative fraction φ_{if} on glassy fibers volume V_{gf} in double logarithmic coordinates for glassy plastics on the basis of polyarylate [64].

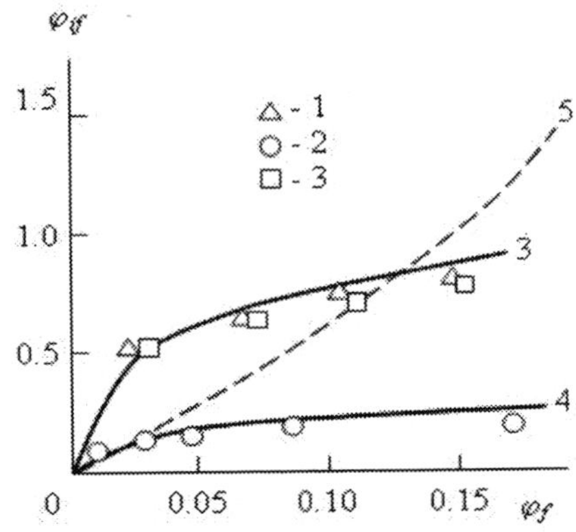

Figure 1.34. The dependence of interfacial layer relative fraction φ_{if} on filler volume contents φ_f for glassy plastics on the basis of polyarylate (1, 3, 6) and PHE-Gr (2, 4, 6). 1, 2 – the calculation according to the equation (1.44); 3, 4, 5 – the calculation according to the equation (1.45) at conditions S_u=variant (4) and S_u=const (3, 5); 6 – the calculation according to the equation (1.11) [69].

Therefore, the fractal analysis methods allow to estimate correctly interfacial layers relative fraction in glassy plastics on the basis of polyarylate as a function of filler volume contents. The continuous approach does not give the correct description of the indicated dependence. Consequently, the stated above results allow to confirm that the only reason of φ_{if} increase at φ_f growth is glassy fiber sizes reduction in composites preparation process owing to its high mechanical brittleness [64].

Another factor, influencing essentially on φ_{if} value in polymer composites, is filler particles aggregation level [24, 39]. The authors [69] fulfilled the estimation of the indicated factor influence degree on the example of polymer composites two series: PHE-Gr and glassy plastics on the basis of polyarylate. As it was noted above, for glassy fiber in virtue of its aggregation absence the value d_{surf}=2.14=const, whereas for composites PHE-Gr filler particles aggregation results to d_{surf} growth from 2.17 up to 2.55 in the range φ_f=0.0088-0.176.

φ_{if} value within the frameworks of continuous approach can be calculated according to the equation [70]:

$$\varphi_{if} = l_{if}\varphi_f S_u \rho_f , \qquad (1.45)$$

where S_u is filler specific surface, ρ_f is its density.

In Figure 1.34 the comparison of values φ_{if} calculated within the frameworks of fractal (the equation (1.44)) and continuous approaches (the equation (1.45)) as function φ_f for glassy plastics on the basis of polyarylate. The good correspondence is obtained for both indicated approaches at using the following values included in the equation (1.45) parameters for glassy fiber: S_u=150 m^2/g and ρ_f=2600 kg/m^3 [65]. In Figure 1.34 the values φ_{if}, calculated according to the equation (1.11), are also adduced, which also well correspond to the obtained ones within the frameworks of the indicated above approaches, i.e. calculated according to the equations (1.44) and (1.45). However, similar calculations for composites PHE-Gr gave essential discrepancy of the results, obtained at continuous and fractal approaches using (Figure 1.34). What is more, continuous approach at φ_f>0.15 gives φ_{if} values >1, that is of no physical significance. Such discrepancy is due to the condition S_u=constant in the equation (1.45) acception. As it is known [65], between S_u values and particles (aggregates of particles) diameter D_{ag} the following relationship exists:

$$S_u = \frac{6}{\rho_f D_{ag}} . \qquad (1.46)$$

From the equation (1.46) it follows that the condition S_u=const does not take into consideration the filler particles aggregation, characterized by D_{ag} growth [71]. As S_u estimations with due regard for filler particles aggregation and PHE chain statistical flexibility change under filler influence [72, 73], S_u value decreases from 265 m^2/g for initial graphite up to 65 m^2/g for filler particles aggregates in composites PHE-Gr with φ_f=0.176. The variable value S_u using in the equation (1.45) gives a good correspondence of φ_{if} values, obtained within the frameworks of continuous and fractal approaches (Figure 1.34).

Let's note an important aspect, connected with the value S_u estimation for filler particles aggregates. The formal fractal analysis supposes S_u growth at d_{surf} increase [74], i.e. at filler particles aggregation intensification [24, 71]. Within the frameworks of fractal analysis S_u values is determined according to the equation [74]:

$$S_u = L^{d_{surf}} r^{2-d_{surf}}, \tag{1.47}$$

where L is fractal linear size, accepted equal to D_{ag}, r is measurement scale, accepted as constant and equal to 5 mcm.

The calculation on the basis of volumes of filler particles aggregate (D_{ag}=40 mcm) and the initial filler particle of diameter 10 mcm at packing coefficient 0.74 [65] shows that one aggregate consists of 59 particles of the initial graphite powder. The value S_u for 59 particles at d_{surf}=2.17 is equal to 3321 relative units, at d_{surf}=2.55 it increases up to 4315 relative units and for filler particles aggregate, consisting of 59 particles, at d_{surf}=2.55 S_u value is equal to 627 relative units. The reasons of aggregate specific surface smaller values in comparison with total S_u value consisting of its particles are obvious: a larger part of these particles surface inside aggregate, is screened by its external surface and consequently can not influence on the interfacial layer formation [69].

Therefore, the data adduced above have shown filler particles aggregation influence on interfacial layer volume in polymer composites. The filler particles aggregation increases aggregate size, decreases its specific surface and reduces interfacial layer relative fraction. For the composites, filled with short fibers, in virtue of application of the indicated above technology of components blending in rotating electromagnetic field this aggregation effect is inhibited. The accounting for aggregation factor allows to obtain a good correspondence of the results, obtained within the frameworks of continuous and fractal approaches [69].

The fractal analysis application for interfacial layers in polymer composites is due to the fact that both filler surface and polymeric matrix are fractal objects [41]. In such treatment interfacial layer can be considered as interaction result of two indicated fractals. As a rule, in polymer composites filler elasticity modulus is essentially higher than polymeric matrix modulus [75] and consequently one should suppose structure modification of polymeric matrix layer, adjoining to filler particles surface and to consider this layer as interfacial one. The authors [76] considered and compared two approaches to interfacial layer description (fractal and continuous) on the example of carbon plastics on the basis of high density polyethylene.

The interfacial layer experimental thickness l_{if} calculation was fulfilled from geometrical considerations as hollow cylinder thickness with internal radius r_{fib} and external one – $r_{fib}+l_{if}$ (that is the equation (1.40) variant) [76]:

$$l_{if} = \left(\frac{\varphi_{if}}{\pi N_{fib} l_{fib}} + r_{fib}^2 \right)^{1/2} - r_{fib}, \tag{1.48}$$

where φ_{if} is the interfacial regions relative fraction, determined according to the equation (1.11), N_{fib} is fibers number per volume unit, l_{fib} and r_{fib} are length and radius of fiber, respectively.

As it is known [77], at two fractal objects interaction the only length scale l exists, defining interpenetration distance of these objects. It is supposed by the indicated above reasons that only for polymer composites filler penetration in polymeric matrix occurs and then $l=l_{if}$ [76]. In this case it can be written [77]:

$$l_{if}^{T} \sim a\left(\frac{r_{fib}}{a}\right)^{2(d-d_{surf})/d}, \qquad (1.49)$$

where index "T" designates l_{if} theoretical value, a is lower linear scale of polymeric matrix fractal behaviour, d is dimension of Euclidean space, in which a fractal is considered (in our case, as before, d=3), d_{surf} – fiber surface dimension.

Allowing for proportionality sign in the relationship (1.49) and the condition r_{fib}=const, the indicated relationship can be written as follows [76]:

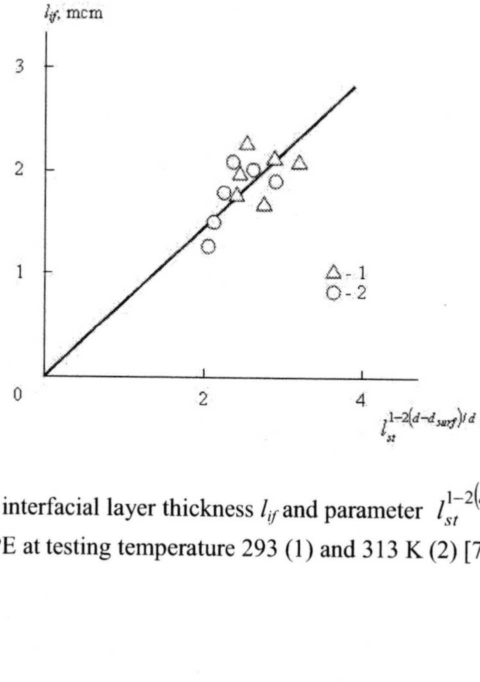

Figure 1.35. The relation of interfacial layer thickness l_{if} and parameter $l_{st}^{1-2(d-d_{surf})/d}$ for carbon plastics on the basis of HDPE at testing temperature 293 (1) and 313 K (2) [76].

$$l_{if}^{T} \sim a^{1-2(d-d_{surf})/d}. \qquad (1.50)$$

For polymeric materials structure as scale a the statistical segment length l_{st} (the equation (1.14)) is accepted.

In Figure 1.35 the relation of parameters l_{if} and $l_{st}^{1-2(d-d_{surf})/d}$, calculated according to the equations (1.48) and (1.14), (1.50), respectively, at d_{surf}=2.14 is adduced. As one can see, between the indicated parameters a good linear correlation is obtained, which is expressed analytically as follows [76]:

$$l_{if} = 0.763 l_{st}^{1-2(d-d_{surf})/d}, \text{ mcm}. \qquad (1.51)$$

Therefore, the data of Figure 1.35 confirm the fractal approach correctness at interfacial layer thickness description and indicate that this layer formation is due to two fractals interaction: filler fibers surface and polymeric matrix structure.

Within the frameworks of continuous approach the value l_{if} calculation can be fulfilled according to the equation (1.43). In Figure 1.36 the dependences $l_{if}(\varphi_f)$, calculated according to the equations (1.43) and (1.51) are adduced. As one can see, the large distinction of continuous conception results (the equation (1.43)) and the experimental data (the equation (1.11)), which is due to the fact, that the indicated conception does not account for polymeric matrix fractal nature. So, at φ_f=0.038 and l_{if}=13.5 mcm, obtained according to the equation (1.43), from the equation (1.51) l_{st}=225 Å will be obtained. Since polymeric matrix structure is a physical fractal within the self-similarity range ~ 5-100 Å [4], then it is obvious, that the formula (1.43) is applicable for Euclidean objects only what one has expected.

Therefore, the stated above results have demonstrated fractal approach correctness at the description of interfacial layer in carbon plastics, which must be considered as two fractals interaction result: filler fibers surface and polymeric matrix structure. Such correctness is achieved only with due regard for matrix polymer molecular characteristics changes, which are due to filler introduction. At the same time, as it was expected, continuous approach did not give correct description of interfacial layer thickness.

Polymers adhesion to solid surfaces is one of the main factors, defining properties of any polymer composites [70]. Many properties of the indicated on composites depend (or are defined by) adhesion level on interfacial boundaries filler-polymeric matrix [78].

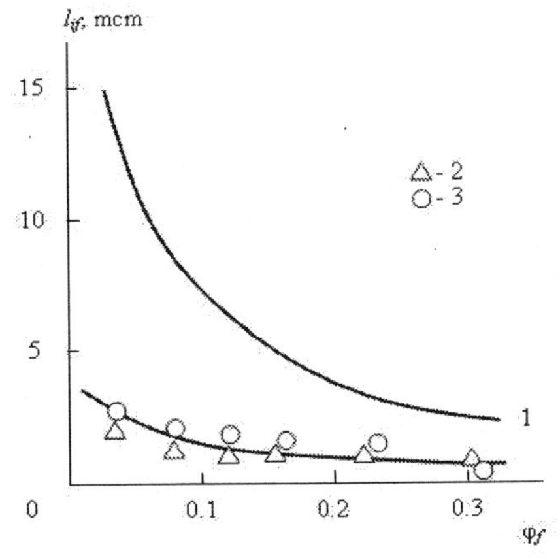

Figure 1.36. The dependences of interfacial layer thickness l_{if} on filler volume contents φ_f calculated according to the equations (1.43) (1), (1.48) (2) and (1.51) (3, 4) for carbon plastics on the basis of HDPE at testing temperatures 293 (3) and 313 K (4) [76].

The adhesion problem is very complex and includes different aspects: chemical, physical and mechanical [70]. Many theoretical approaches exist to describe and explain adhesion phenomena. However, not a single existent theory gives the possibility to calculate adhesional interaction energy and adhesional junction strength. This is due to the fact, that a large number of heterogeneous factors influence, on adhesion actually and on adhesional strength which can not be accounted within the frameworks of the only one theory [70]. The authors [75, 79] on the example of particulate-filled composites PHE-Gr have shown the availability of structural aspect of interfacial adhesion, on the level of which two factors, as minimum, influence: the level of physical and/or chemical interactions polymer-filler and interfacial layer structure.

As it was shown above [57], modern studies in surface chemistry and physics field allow to describe the processes, connected with surface energy role in structure and properties formation of nanometer scale substance. In addition a structure universal informant is fractal dimension [54]. On the basis of this conception the authors [80] studied interfacial adhesion structural aspect in polymer composites, filled with short fibers. The surface energy value in paper [80] was defined by difference $d- d_f^{if}$, where d=3 and d_f^{if} varies within the limits of $2 \le d_f^{if} < 3$ [55].

In Figure 1.37 the dependence of interfacial layer strength σ_a on dimensions difference $(d- d_f^{if})$ for carbon plastics on the basis of phenylone is adduced. As one can see, the value d_f^{if} decrease or difference $(d- d_f^{if})$ increase the results to interfacial layer strength growth in the range ~ 10-69 MPa. This problem energetic aspect can be estimated by using energy density W_p, "pumping" in interfacial layer at its mechanical loading [41]. W_p value is determined according to the equation [41]:

$$W_p = \frac{\sigma_a^2}{2E_{if}}, \tag{1.52}$$

where E_{if} is interfacial layer elasticity modulus, the value of which can be estimated as follows [4]:

$$E_{if} = 0.7\left(\frac{S}{C_\infty}\right)^{1/2}, \text{ GPa.} \tag{1.53}$$

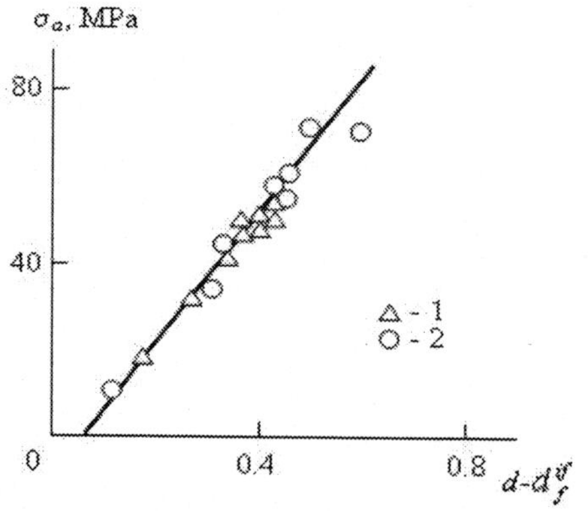

Figure 1.37. The dependence of interfacial layer strength σ_a on dimensions difference $(d- d_f^{if})$ for carbon plastics on the basis of phenylone, obtained with magnetic (1) and mechanical (2) separation application [80].

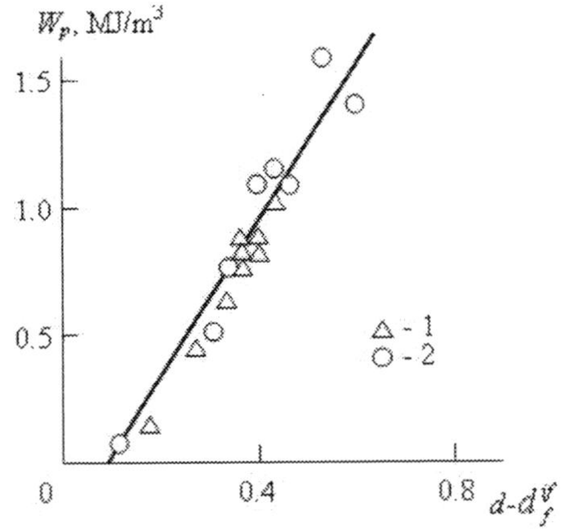

Figure 1.38. The dependence of energy density W_p on dimensions difference (d- d_f^{if}) for carbon plastics on the basis of phenylone, obtained with magnetic (1) and mechanical (2) separation application [80].

In Figure 1.38 the dependence of W_p on dimensions difference (d- d_f^{if}) is shown, from which the linear growth W_p at (d- d_f^{if}) increase follows. Since the value E_{if}=const, then from the data of Figure 1.38 it follows, that decreasing of interfacial layer substance distribution dimension d_f^{if} in three-dimensional space results to energy release, which at mechanical loading is required to be compensated by "pumping" mechanical energy W_p, after which interfacial layer failure will occur [80].

Let's note an interesting aspect, following from the comparison of Figs. 1.31 and 1.37 graphs. The difference (d_f^{if} -d_f), independently from these dimensions absolute values, results to σ_a rising or W_p increasing. Therefore, for interfacial layer strength increase and, consequently, macroscopic strength of composite, it is necessary to form interfacial layer as much as possible structurally close to polymeric matrix and $d_f^{if} \to 2$.

In paper [81] it was offered to characterize polymer-filler interaction degree (adhesion level) with the aid of parameter A, the value of which is determined according to the equation:

$$A = \frac{1}{1-\varphi_f} \cdot \frac{tg\delta^c}{tg\delta^p} - 1, \qquad (1.54)$$

where tg δ^c and tg δ^p are tangents of mechanical (dielectric) losses angle for composite and matrix polymer, respectively.

Strong interactions between filler and polymer matrix on interfacial border try to reduce molecular mobility in filler surface locality in comparison with bulk polymeric matrix. This results to tg δ^c decrease and, consequently, A. Therefore, the small value A indicates to high interaction degree or adhesion between polymer composite phases [81].

In Figure 1.39 the dependence of parameter A on dimensions difference (d- d_f^{if}) for particulate-filled composites PHE-Gr, is adduced plotted according to the data of paper [75]. As one can see, the fast A reduction or interfacial adhesion intensification at (d- d_f^{if}) growth is observed and at (d- d_f^{if})≈0.39 the value A reaches its smallest value: A=-1, that according to the equation (1.54) corresponds to the condition tg δ^c=0 or molecular mobility "freezing". In this case for (d- d_f^{if})>0.39 A=-1=const was accepted. Further the plot of Figure 1.39 can be used as a calibrating one for the studied carbon plastics. In Figure 1.40 the generalized dependence $\sigma_a(A)$ for carbon plastics and composites PHE-Gr is shown, from which σ_a growth at A reduction or interfacial adhesion intensification for both types of composites follows. At A=-1 the value σ_a for carbon plastics grows at constant A value, that is due to interfacial layer packing density increase in virtue of d_f^{if} decreasing or difference (d- d_f^{if}) rising above the value 0.39, i.e. by purely structural causes. Let's note, that A=-1 is reached at bulk polymeric matrix structure fractal dimension d_f, which is approximately equal to $d_f \approx d_f^{if}$ ≈2.50. As it is known, the indicated dimension corresponds to Witten-Sander cluster structure, by a set of which polymers amorphous state structure is simulated [82, 83]. This dimension reaching means chains parts stretching between clusters and, as consequence, their molecular mobility loss [83], i.e. the achievement of the condition tg δ^c=0. It is significant that this condition is realized at $d_f \approx d_f^{if}$, i.e. when interfacial layer structure is not distinguished from the bulk polymeric matrix structure. This has been expected, since tg δ^c characterizes not interfacial layer actually, but the entire composite.

From the equation (1.54) it follows that the condition A=0 at small φ_f is reached at approximate equality tg $\delta^c \approx$ tg δ^p or at close values of adhesional and cohesive interactions. Hence, a negative A values correspond to the situation, where adhesional interactions are stronger than cohesive ones, that increases the interfacial layer strength σ_a and, consequently, the entire composite strength [80].

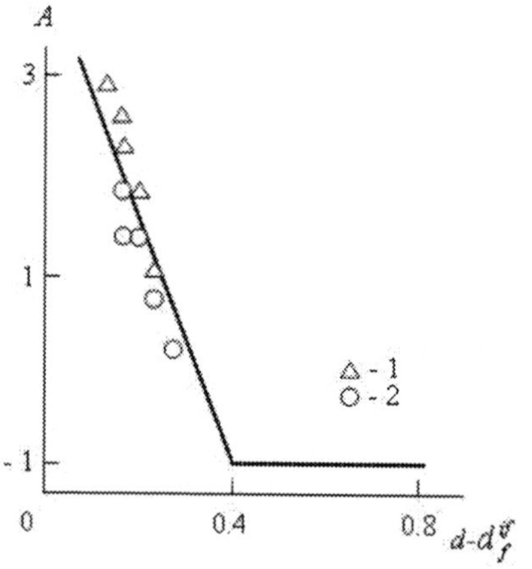

Figure 1.39. The dependence of parameter A on dimensions difference (d- d_f^{if}) for composites PHE-Gr-I (1) and PHE-Gr-II (2) [80].

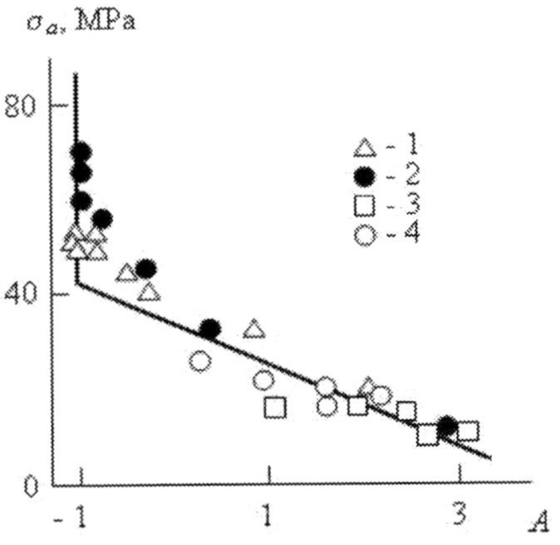

Figure 1.40. The dependence of interfacial layer strength σ_a on parameter A for carbon plastics on the basis of phenylone, obtained with magnetic (1) and mechanical (2) separation application and composites PHE-Gr-I (3) and PHE-Gr-II (4) [80].

Therefore, the adduced above results have demonstrated that for the composites, filled with short fibers, interfacial layer strength is defined by both polymer-filler physical and/or chemical interactions level and interfacial layer structure state similarly to particulate-filled composites. At certain conditions adhesional interactions on interfacial boundary can exceed cohesive interactions in a bulk polymer. The greatest adhesion level is realized at the same fractal dimensions of structure of interfacial layer and bulk polymeric matrix [80].

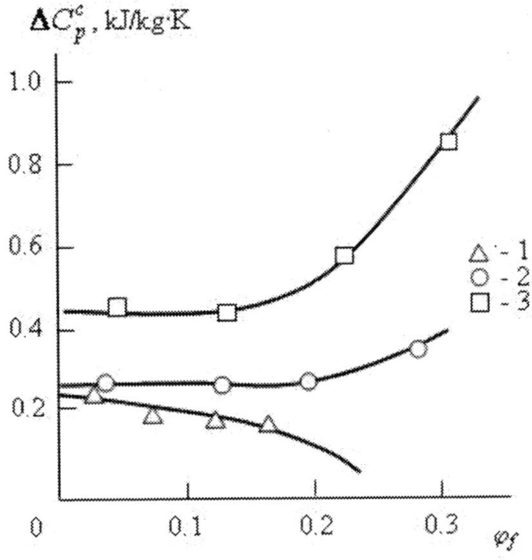

Figure 1.41. The dependences of heat capacity at constant pressure jump at glass transition temperature ΔC_p^c on fibers volume contents φ_f for composites PAr-glassy fiber (1), PAr-uglen (2) and PAr-vniivlon (3). The horizontal shaded line shows the ΔC_p^p value for PAr [85].

Polymers filling with organic fibers has its special features and one of them is possibility of chemical bonds between polymeric matrix and filler formation [84]. This effect changes interfacial layers polymer-filler structure and properties and is an important influence factor on the entire composite. The study of polyarylate (PAr), filled with short fibers, by IR-spectroscopy methods experimentally confirms chemical bonds between PAr matrix and fibers vniivlon and uglen formation and their absence in case of glassy fibers [25]. This binding and filler interaction is mainly contained in intermolecular bonds arising at the expence of breaking away from quartering carbon atom of diphenylpropane methyl groups and iminne groups formation. The authors [85] studied polymeric matrix-fiber physical and/or chemical bonds formation influence on structure and properties of composites on the basis of PAr, filled with fibers of two types: organic (uglen, vniivlon) and glassy fibers [25].

As it was noted above, for fillers influence study the calorimetric method is successfully applied, allowing to determine a number of important characteristics of filled polymeric systems [30]. The experimental data show on the whole the decrease of heat capacity at constant pressure jump ΔC_p^c at glass transition temperature at filler volume contents φ_f growth. This testifies unequivocally to macromolecules some part exclusion from participation in vitrification cooperative process owing to its interaction with filler surface [30]. In Figure 1.41 the dependence of ΔC_p^c on φ_f for three studied composites is adduced. As one can see, for various fillers the function ΔC_p^c (φ_f) different behaviour is observed. For composites PAr-vniivlon the sharp increase ΔC_p^c at φ_f growth is obtained, being of small φ_f of order 0.10 the heat capacity jump value for composite ΔC_p^c exceeds this parameter for initial polymer ΔC_p^P (horizontal shaded line in Figure 1.41). For composites PAr-glassy fibers the dependence ΔC_p^c (φ_f) has an expected character – ΔC_p^c value reduces from 0.24 up to 0.10 kJ/(kg·K) at φ_f increase from 0.024 up to 0.199. Such dependence ΔC_p^c (φ_f) assumes interfacial regions relative fraction φ_{if} increase, which can be calculated according to the equation (1.11) at ΔC_p^P=0.49 kJ/(kg·K) [25]. The calculation according to this equation shows φ_{if} increase from 0.51 up to 0.80 at φ_f growth from 0.024 up to 0.199 (see Figure 1.34).

For composites PAr-uglen the dependence ΔC_p^c (φ_f) occupies intermediate location between similar curves for the two considered above composites on the basis of PAr. The increase of ΔC_p^c at φ_f growth for composites PAr-vniivlon can be explained within the frameworks of qualitative model, offered by Lipatov [30]. If in correspondence to notions about adsorbed macromolecules conformations to consider that macromolecule on the surface assumes loop form, but does not stretch on it, then it is quite obvious that such loop ends are connected with one another essentially weaker than in bulk. Hence, oscillations amplitude of chain parts between points of contacts with surface increases that results to heat capacity absolute value rising. In addition (since loops size is probably smaller than chain statistical segment length) segmental mobility can be decreased (ΔC_p^c reduction), whereas oscillations amplitude of separate atomic groups in loop will be increased (hence ΔC_p^c rising) [30].

Therefore, the comparison of ΔC_p^c (φ_f) data for composites PAr-glassy fiber and PAr-vniivlon demonstrates that for the first polymer macromolecules stretching on the fiber surface and strong physical interactions inhibit molecular mobility in interfacial layer and for the second pointed chemical bonds and the absence of the physical ones intensifies it.

The other explanation of ΔC_p^c growth for composites PAr-vniivlon can be given within the frameworks of fractal analysis. According to the mentioned above Bunderlich rule heat capacity jump at vitrification ΔC_p^c at calculation per one mole of macromolecule minimum (by sizes) structural elements is a constant value, which is equal to 0.173-0.122 kJ/(kg·K) and is calculated according to the formula (1.34), from which $\overline{M}$ decrease and, hence, the indicated minimum structural element (statistical segment) length reduction at φ_f growth and corresponding ΔC_p^c increasing (Figure 1.41) follow. In its turn, in virtue of l_0=const, l_s decrease defines C_∞ reduction. The interconnection of C_∞ and molecular mobility degree, characterized by fractal dimension D_{ch} of chain part between chemical bonds points, is given as follows [86]:

$$\frac{2}{\varphi_{cl}} = C_\infty^{D_{ch}} . \tag{1.55}$$

From the equation (1.55) D_{ch} growth at C_∞ decrease (C_∞), molecular mobility intensification and, consequently, ΔC_p^c increase follows. Let's note, that similar C_∞ reduction has been observed for cross-linked epoxy polymers at cross-linking degree increase [87]. Therefore, all the said above supposes that between PAr polymeric matrix and vniivlon fibers dense enough network of chemical bonds is formed.

For composites PAr-glassy fiber polymer macromolecules stretching should be expected on these fibers relatively smooth surface and, as consequence, interfacial layers formation, connected with filler by physical bonds [31]. Composites PAr-uglen in virtue of the dependence ΔC_p^c (φ_f) intermediate position obviously form composite intermediate structure with chemical bonds limited number and φ_{if} small value, although one should expect that the stronger chemical bonds formation is expressed. So, calculation according to the equation (1.11) shows φ_{if} reduction from 0.45 up to 0.29 in the range φ_f=0.038-0.282 [85].

The dependence of ΔC_p^c on the considered composites structural characteristics can be traced from the data of Figure 1.42, where the correlations of their desities difference, determined experimentally (ρ_e) and calculated according to the additive scheme (ρ_{ad}): $\Delta\rho=\rho_e-\rho_{ad}$, and ΔC_p^c value. It is interesting that these data suppose ΔC_p^c ($\Delta\rho$) correlation, which is opposite to the expected one. So, for composites PAr-vniivlon and PAr-uglen $\Delta\rho>0$, i.e. some structure densification is observed, from which one should expect molecular mobility degree reduction. Nevertheless, ΔC_p^c growth in this case assumes that the described above C_∞ reduction compensates completely $\Delta\rho$ increase. For composites PAr-glassy fiber $\Delta\rho<0$, i.e.

structure loosening is observed, accompanied, nevertheless, by ΔC_p^c reduction. This circumstance indicates that molecular mobility growth is strongly limited by physical bonds on interfacial boundary polymer-filler and this limitation influence on the entire composites structure is defined by large value φ_{if}=0.51-0.80 [85].

And at last the authors [85] traced the considered above bonds polymer-filler influence on composites properties on the example of their elasticity modulus E. In Figure 1.43 the dependences of E on filler mass contents W_f for the three considered composites is shown. The data of Figure 1.43 showed fast growth E at W_f increase for PAr-vniivlon and PAr-glassy fiber and much slower – for PAr-uglen. Hence, these data indicate that elasticity modulus increase requires strong bonds on polymer-filler boundary and confirm the conclusion made by Lipatov [84] – the problem consists not so much in nature as in strong bonds quantity, which are necessary for optimal properties ensuring. Both physical bonds in composites PAr-glassy fiber and chemical bonds in composites PAr-vniivlon satisfy to these requirements [85].

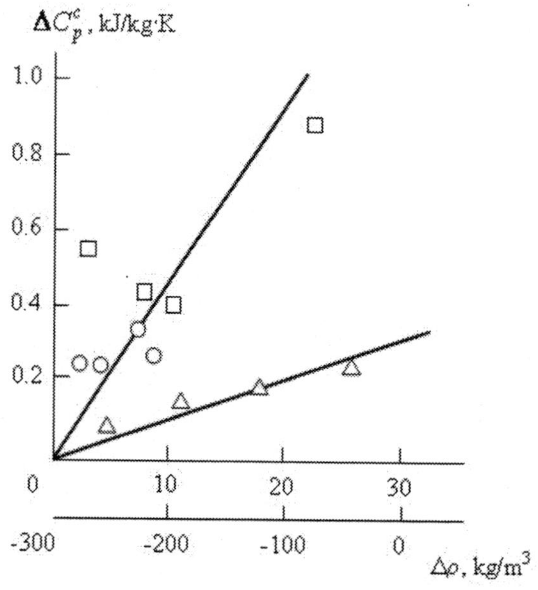

Figure 1.42. The dependences of heat capacity at constant pressure jump at glass transition temperature ΔC_p^c on experimental and additive density difference Δρ for composites on the basis of PAr. The designations are the same as in Figure 1.41. The upper axis of abscisses – for PAr-uglen and PAr-vniivlon, lower one – for PAr-glassy fiber [85].

Therefore, the adduced above results allow to identity bonds type on polymer-fiber interfacial boundary. For the organic fibers vniivlon on the basis of rigid-chain polyheterarylene chemical bonds are the main type and for glassy fiber – the physical ones. In this respect uglen occupies intermediate position with some chemical bonds prevalence. For reinforcement (elasticity modulus increase) of composites nature is of no importance, but strong bonds quality is, which is necessary for this effect realization.

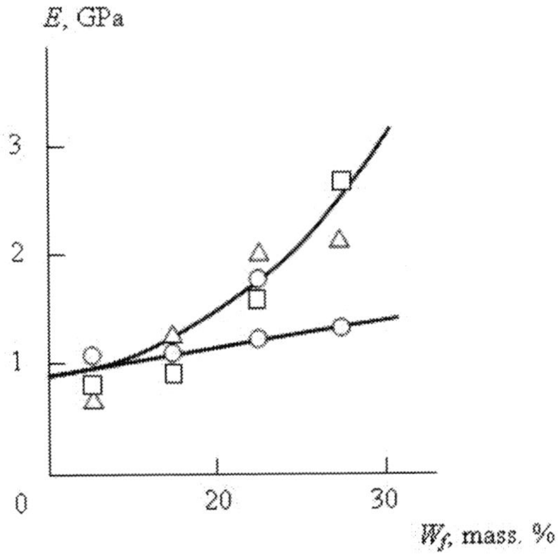

Figure 1.43. The dependences of elasticity modulus E on fibers mass contents W_f for composites on the basis of PAr. The designations are the same as in Figure 1.41 [85].

As it is known [22, 23], one of the most important factors, influencing on strength of polymer composites, filled with short fibers is fibers orientation degree of polymer matrix. The offered by the authors [9] components "dry" blending method in rotating electromagnetic field with the aid of nonequiaxial ferromagnetic particles assumes this factor change, and, consequently, composites strength change. However, composites are structurally complex solids, including interfacial layers polymer-filler. Interfacial interactions define to a considerable extent the indicated layers structure and properties, influencing thereby on the entire complex of polymer composites properties [84]. As it was shown above between filler and polymer interactions of different kinds could exist – both physical and chemical ones [85]. Proceeding from this, the authors of paper [88] fulfilled the research of interfacial bonds type influence on filler fibers orientation degree (and, consequently, on strength) in composites on the basis of PAr.

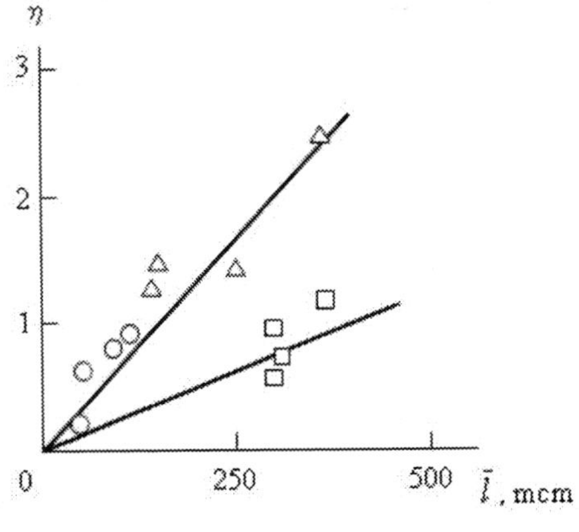

Figure 1.44. The dependences of orientation factor η on fiber average length $\bar{l}$ for composites on the basis of PAr. The designations are the same as in Figure 1.41 [88].

A ferromagnetic particles rotation in rotating electromagnetic field should result to fibers orientation improvement, which can be characterized by the orientation factor η, determined with the aid of the equation (1.6). In Figure 1.44 the dependence of calculated by the indicated mode η on fiber average length $\bar{l}$ for the three studied composites on the basis of PAr is adduced. $\bar{l}$ change is due to reduction of fragments of initial brittle fiber during its processing [25, 58]. As one can see, the linear dependence $\eta(\bar{l})$ is obtained, which breaks up into two sections: with a higher slope (for Par-uglen and PAr-glassy fiber) and smaller slope one (for PAr-vniivlon). Since, as it was shown above, the most intensive chemical bonds are formed just in composites PAr-vniivlon, then the conclusion can be made, that the indicated bonds most effectively violate fibers orientation, obtained as a result of components preliminary blending in rotating electromagnetic field. Physical bonds formation violates fibers preliminary orientation much weaker or does not violate at all as it follows from the data of Figs. 1.41 and 1.44 comparison.

As it was shown above, for the considered composites ΔC_p^c reduction means physical bonds intensification on interfacial boundary polymer-filler and this parameter increasing means chemical bonds intensification. Therefore the value ΔC_p^c can be used for normalization of the dependences $\eta(\bar{l})$ shown in Figure 1.44. As it follows from the data of Figure 1.45, the dependence η on generalized parameter $\bar{l}/\Delta C_p^c$ is approximated actually by the only linear correlation for the three studied systems, which is expessed analytically as follows [88]:

$$\eta = 1.4\times10^{-3}\frac{\bar{l}}{\Delta C_p^c}, \qquad (1.56)$$

where $\bar{l}$ is given in mcm, and ΔC_p^c - in kJ/(kg·K).

The obtained in Figure 1.45 general correlation confirms the proposal done above, that chemical bonds polymer-filler violate fibers preliminary orientation in polymeric matrix much stronger than physical ones. This can be explained as follows. Chemical bonds are formed in melt at temperature ~ 523 K, which contributes to chemical reactions proceeding. Since these bonds are strong enough and polymer melt viscosity is essentially lower than solid polymer, then chemical bonds dense enough network can result to fibers initial orientation violation. Nevertheless, if these bonds density is small, as in case of PAr-uglen system, then melt high enough viscosity does not allow fibers preliminary orientation changing. As for PAr-glassy fiber system, then interfacial bonds are formed in it by macromolecules stretching on filler smooth enough surface and strong enough physical bonds are formed only in glassy state, where fibers mobility is inhibited completely.

Therefore, from the equation (1.6) it follows that at each fixed value φ_f two factors, capable to influence in preparation process on strength of the final article from composites: orientation factor η change owing to interfacial bonds formation and ($\bar{l}/\bar{d}$) variation owing to filler fibers crushing. The data of Figure 1.44 suppose that for PAr-vniivlon system the first factor has decisive significance and for PAr-glassy fiber – the second one, since $\bar{l}$ decrease for it is expressed much stronger than for the first from the indicated systems. In Figure 1.46 the dependence of the determined experimentally strength values σ_f^c on fibers volume contents φ_f for the two mentioned systems is adduced.

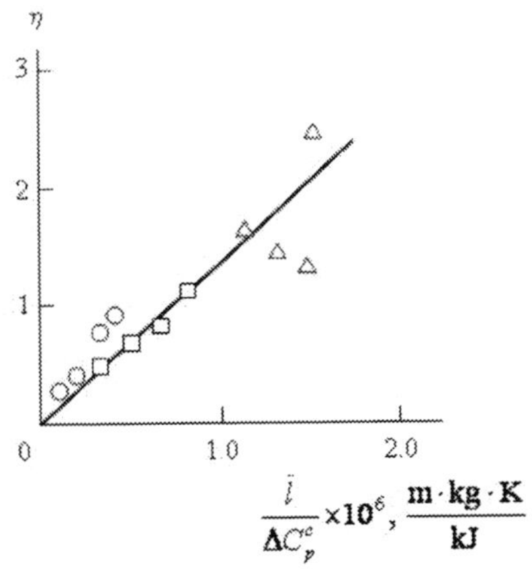

Figure 1.45. The dependence of orientation factor η on generalized parameter ($\bar{l} / \Delta C_p^c$) for composites on the basis of PAr. The designations are the same as in Figure 1.41 [88].

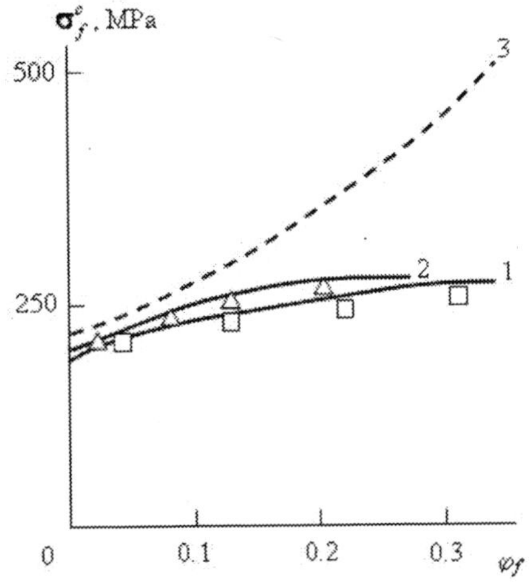

Figure 1.46. The dependences of experimentally determined fracture stress σ_f^c on filler volume contents φ_f for PAr-vniivlon (1) and PAr-glassy fiber (2). 3 – the theoretical dependence σ_f^c (φ_f), calculated according to the equation (1.6). The explanation is given in the text [88].

As one can see, in virtue of the considered above factors influence the value σ_f^c weak increase with approximately the same of its absolute magnitudes for both composites at φ_f growth is observed. Further the authors [88] have calculated the theoretical dependence σ_f^c (φ_f) according to the equation (1.6) at the condition that in PAr-vniivlon system the value η is the same as in PAr-glassy fiber, and in the last one fiber reduction to fragments is the same, as in PAr-vniivlon system. The theoretical dependences σ_f^c (φ_f), shown by shaded line in Figure 1.46, are approximately the same for the considered composites and found essential increase of σ_f^c in comparison with the experimental data. This comparison clearly demonstrates strength increase possible directions for composites filled with short fibers.

Therefore, the stated above results showed interfacial bonds type influence on filler fibers orientation degree in composites, prepared by technology of components preliminary blending in rotating electromagnetic field. Strong chemical bonds polymer-filler violate this orientation in polymer melt and physical bonds do not practically influenced on it. The essential reserve of the considered composites strength increasing is weakening of fiber reduction to fragments and its orientation initial degree preservation [88].

Within the frameworks of fractal analysis the essential influence on interfacial regions structure of both filler surface structure and their formation mechanism, which is connected with participation extent of diffusive processes in the indicated regions formation [61, 62, 89]. In the general case this interconnection can be formulated as follows: the larger diffusive processes role in interfacial layer formation the higher interpenetration of macromolecular coils in it and the higher its strength [61]. Therefore the authors of paper [90] studied a diffusive processes role in interfacial layers formation on the example of carbon plastics on the basis of phenylone with fractional derivatives theory and fractal analysis enlisting.

At present the division of diffusive processes on the slow and fast ones is generally accepted [91, 92]. In such division basis the dependence of a diffusible particle displacement s on time t is put [92]:

$$s = t^{\beta}, \qquad (1.57)$$

where for the classical case β=1/2, for slow diffusion β<1/2 and for fast one - β>1/2.

Within the frameworks of fractional derivation theory it was shown that its main parameter – fractional exponent α is connected with both β and structure characteristic – fractal dimension by different functional forms depending on diffusion type. In other words, structures with the same d_f can have very differing diffusivities D. The interconnection of d_f (in our case – interfacial layer structure fractal dimension d_f^{if}) and α for three-dimensional Euclidean space can be obtained by analogy with paper [92] as follows. Let's suppose that Hurst exponent H is connected with d_f by the equation [90]:

$$d_f = 3 - H. \qquad (1.58)$$

The authors [92] have shown that in the relationship (1.58) the exponent β is equal to (1-α)/2 for slow diffusion and 1/(1+α) for fast one. Equaling H to the indicated expressions for α according to the methodics [11], let's obtain the interconnection of α and d_f^{if} [90]:

$$\alpha = \frac{3 - d_f^{if}}{2} \qquad (1.59)$$

for slow diffusion and

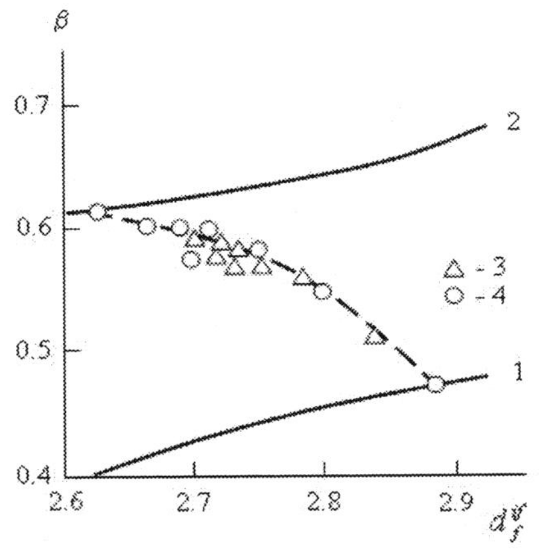

Figure 1.47. The dependence of exponent β on interfacial layer structure fractal dimension d_f^{if} for carbon plastics on the basis of phenylone. Theoretical calculation for slow (1) and fast (2) diffusion; calculation according to the equation (1.57) for samples, prepared with magnetic (1) and mechanical (2) separation application [90].

$$\alpha = \frac{1}{d_f^{if} - 1} \tag{1.60}$$

for fast diffusion.

The second exponent α calculation method consists in the relationship (1.57) direct application at the condition that interfacial layer thickness l_{if} is equal to s and the value t is accepted equal to composite processing duration. In the considered case t=300 s and the value l_{if} was calculated according to the equation (1.48). In Figure 1.47 the comparison of the obtained theoretically and calculated according to the equation (1.57) dependences of β on d_f^{if} is adduced. As one can see, β values, calculated according to the equation (1.57), i.e. according to the experimental l_{if} and t magnitudes, are intermediate ones between the theoretical dependences β(d_f^{if}) for slow and fast diffusion. d_f^{if} increase results to β reduction and defines transition from fast diffusion to slow one. In its turn, this results to l_{if} decrease from 2.77 up to 1.18 mcm.

In order to elucidate physical grounds of this effect the authors [90] fulfilled calculation of gyration radius $\langle R_g^2 \rangle^{1/2}$ of macromolecular coil in interfacial layer. Within the frameworks of fractional derivatives theory this parameter can be determined according to the equation [92]:

$$\langle R_g^2 \rangle^{1/2} = \frac{l_{st}}{(1-\alpha)\Gamma(\alpha)} N^{1-\alpha}, \tag{1.61}$$

where l_{st} is statistical segment length, $\Gamma(\alpha)$ is Eiler gamma function, N is polymerization degree.

The interconnection between interfacial layer condensed state structure, characterizing by dimension d_f^{if}, and macromolecular coil fractal dimension D_c for linear polymers is given by the equation (1.30) and fractional exponent α value for such coil is determined as follows [93]:

$$\alpha = D_c - D_c^l, \tag{1.62}$$

where D_c^l is fractal dimension of "leaky" coil, which is equal to 1.50 [51].

Eiler gamma function for the variable α has the following form [10]:

$$\Gamma(\alpha) = \left(\frac{\pi}{2}\right)^{1/2} \alpha^{\alpha} e^{-\alpha}, \tag{1.63}$$

and the value N is accepted equal to 200.

In Figure 1.48 the dependence of β value, calculated according to the equation (1.57), on macromolecular coil gyration radius $\langle R_g^2 \rangle^{1/2}$ is adduced. As it follows from the data of this Figure, $\langle R_g^2 \rangle^{1/2}$ increase or coil stretching degree rise on fiber surface results to diffusive processes weakening expressed by β reduction. In other words the larger coil is stretched on fiber surface the denser interfacial layer structure and the higher its fractal dimension d_f^{if}, that results to macromolecular coils interpenetration weaking diffusive processes inhibition and β reduction, in the long run defining transition from the fast diffusion to the slow one [90].

Let's consider diffusive processes influence on interfacial layer strength σ_a, determined according to the equation (1.7). In Figure 1.49 the dependence of σ_a on exponent β, calculated according to the equation (1.57), is shown, from which σ_a fast growth at β increase follows. Hence, diffusive processes role intensification at interfacial layers formation contributes to their strength increase. The physical ground of this effect is obvious – diffusion intensification contributes to macromolecular coils interpenetration, denser physical entanglements network formation, that increases interfacial layer strength [61].

Therefore, the stated above results showed the interfacial layers structure influence on diffusive processes intensity, which proceed at their formation. The increase of these layers structure fractal dimension, which is due to macromolecular coils stretching on fiber smooth surface, defines transition from the fast diffusion to the slow one. This factor controls both interfacial layer thickness and its strength [90].

As it was shown above, interfacial layers formation on boundary polymer-filler in composites, filled with short fibers, has its specific features. These fibers possess smooth enough surface, i.e. its fractal dimension d_{surf} is close to 2 ($2.0 \le d_{surf} < 3$) [31]. This circumstance supposes polymer chains conformations change in interfacial layer, which is expressed in these chains stretching on fiber surface [31]. Consequently interfacial layers structure in such composites differs from the bulk polymeric matrix structure. The specific feature of macromolecules on division boundary polymer-fiber is macromolecular coil anisotropy, which should be simulated not by a sphere, but by a stretched rotation ellipsoid [94]. To describe this anisotropy degree is possible within mathematical calculus of fractional integration and differentiation, which was used successfully for coil structure characterization in diluted solution [92, 95]. In paper [96] such analysis was made for interfacial layers of carbon plastics on the basis of phenylone.

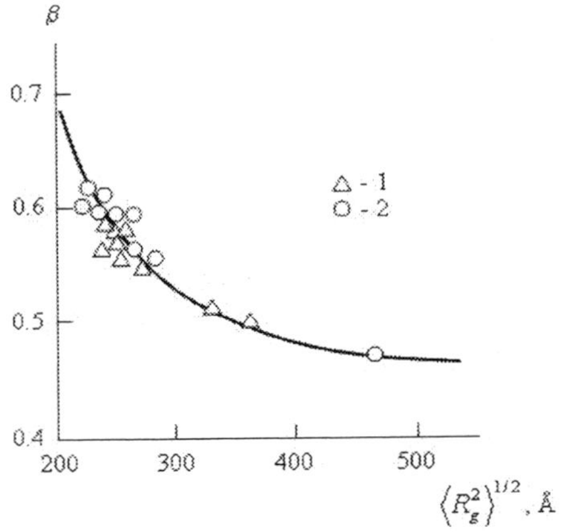

Figure 1.48. The dependence of exponent β on macromolecular coil gyration radius $\langle R_g^2 \rangle^{1/2}$ in interfacial layer for carbon plastics on the basis of phenylone, prepared with magnetic (1) and mechanical (2) separation application [90].

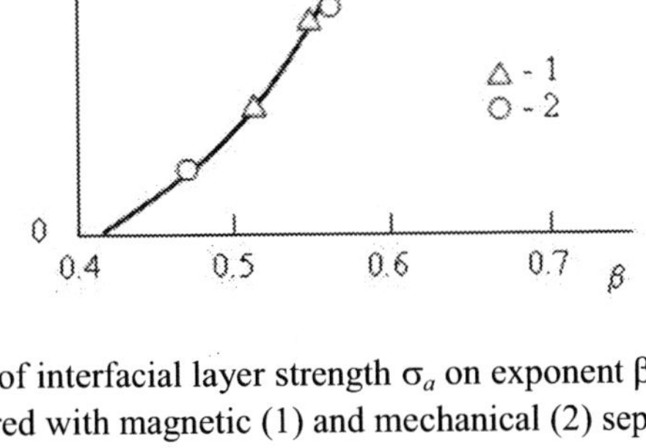

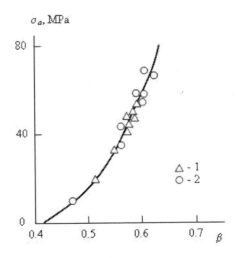

Figure 1.49. The dependence of interfacial layer strength σ_a on exponent β value for carbon plastics on the basis of phenylone, prepared with magnetic (1) and mechanical (2) separation application [90].

Within the frameworks of fractional derivatives theory the macromolecular coil gyration radius $\left\langle R_g^2 \right\rangle^{1/2}$ value can be determined according to the equation (1.61). In Figure 1.50 the dependence of calculated by the indicated mode value $\left\langle R_g^2 \right\rangle^{1/2}$ on interfacial regions relative fraction φ_{if} for carbon plastics on the basis of phenylone. As one can see, the absolute value $\left\langle R_g^2 \right\rangle^{1/2}$, characterizing macromolecular coil anisotropy degree at its stretching on fiber surface, reduces fastly at φ_{if} growth and at $\varphi_{if} \approx 0.40$ tries to attain asymptotically $\left\langle R_g^2 \right\rangle^{1/2}$ value for isotropic coil (horizontal shaded line). This tendency indicates that the mentioned macromolecules stretching is effective only for chains, finding in direct proximity from fiber surface and the remaining macromolecules in interfacial layer try to preserve statistical coil conformation.

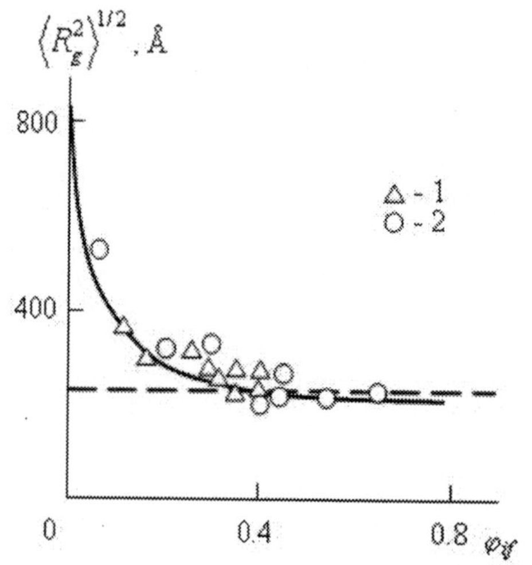

Figure 1.50. The dependence of macromolecular coil gyration radius $\left\langle R_g^2 \right\rangle^{1/2}$ in interfacial layer on its relative fraction φ_{if} for carbon plastics on the basis of phenylone, prepared with magnetic (1) and mechanical (2) separation application [96].

Let's consider interfacial layers formation mechanism in the studied carbon plastics. Meakin shows Witten-Sander model applicability for simulation of depositions on fibers and surfaces [97, 98] and further this conception was successfully used for interfacial layers formation mechanism description in particulate-filled composites [62, 83, 89, 99]. In paper [98] the following relationship between mean-square thickness of interfacial layer l_{if} and particles number n_i in it was obtained:

$$l_{if} \sim n_i^{\varepsilon}, \tag{1.64}$$

where exponent ε varies from 1.7 for depositions, controlled by diffusion, up to 1.0 for conditions, where diffusive processes are unessential.

ε value can be determined from the following equation [98]:

$$\varepsilon = \frac{1}{1 - d + d_f^{if}}, \tag{1.65}$$

where d is dimension of Euclidean space, in which a fractal is considered (it is obvious, that in the considered case d=3).

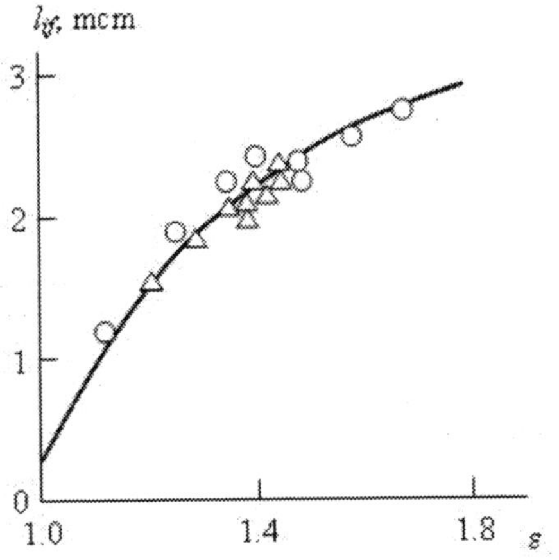

Figure 1.51. The dependence of interfacial layer thickness l_{if} on exponent ε. The designations are the same as in Figure 1.50 [96].

In Figure 1.51 the dependence l_{if} on ε value is shown, where l_{if} is determined according to the equation (1.48). As one can see, ε increasing or diffusive processes intensification results to l_{if} growth and, hence, φ_{if}. It is significant that at ε=1.0 the dependence $l_{if}(\varepsilon)$ extrapolates to $l_{if}\approx 0.25$ mcm, i.e. interfacial layer thickness, formed only by macromolecules stretching on fiber surface without diffusive processes participation, is very small.

The macromolecular coils number in interfacial layer n_{if} can be calculated according to the following simple formula [96]:

$$n_{if} = \frac{l_{if}}{\left\langle R_g^2 \right\rangle_{\perp}^{1/2}} \tag{1.66}$$

where $\left\langle R_g^2 \right\rangle_{\perp}^{1/2}$ is macromolecular coil gyration radius in direction perpendicular to fiber surface, which can be easily determined from purely geometrical considerations.

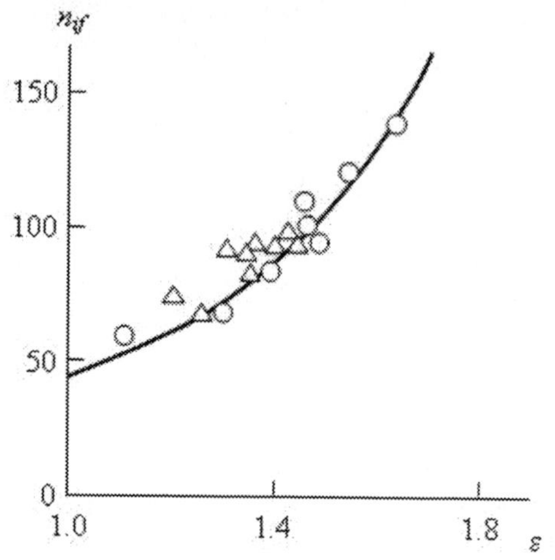

Figure 1.52. The dependence of macromolecular coils number n_{if} in interfacial layer on exponent ε. The designations are the same as in Figure 1.50 [96].

In Figure 1.52 the dependence n_{if} on exponent ε is adduced, from which n_{if} increasing at ε growth follows. In other words, diffusive processes intensification at formation of carbon plastics interfacial layers increases macromolecular coils number in them. It is significant that at diffusive processes absence, i.e. at ε=0, n_{if} value is finite and equal to ~ 44. If to account that this condition defines the smallest value $\left\langle R_g^2 \right\rangle_{\perp}^{1/2}$ ≈50 Å too, then chains stretching on fiber surface mechanism, proceeding without diffusive processes participation, gives the smallest l_{if} magnitude, which is equal to ~ 0.22 mcm, that corresponds well to the data of Figure 1.51.

The authors [94] obtained the equation for macromolecular coil compression maximum degree λ^{comp} ($\lambda^{comp}=\left\langle R_g^2 \right\rangle^{1/2} / \left\langle R_g^2 \right\rangle_{\perp}^{1/2}$) in the following form:

$$\lambda^{comp} = N^{-(D_c-3)/6}. \tag{1.67}$$

In Figure 1.53 the dependence of $n_{if}^{1/3}$ on λ^{comp} for the considered carbon plastics is shown, which is well approximated by the linear correlation. The indicated correlation describes compression maximally possible degree, which for N=const is defined by coil fractal dimension D_c. D_c increase or coil in interfacial layer compactness rise results to λ^{comp} reduction, corresponding n_{if} decrease and, consequently, l_{if} and φ_{if}. Let's estimate quantitatively this effect. The relation between n_{if} and λ^{comp} can be expressed analytically as follows [96]:

$$n_{if} = 14\left(\lambda^{comp} - 1\right). \tag{1.68}$$

Let's obtain further from the equation (1.67) that D_c increase from 1.50 up to 2.0 results to λ^{comp} reduction from 3.76 up to 2.42 and according to the equation (1.68) let's obtain n_{if} decrease from ~ 300 up to ~ 40, i.e. almost on one order of value. The adduced example demonstrates macromolecular coil structure importance in interfacial layers of carbon plastics formation.

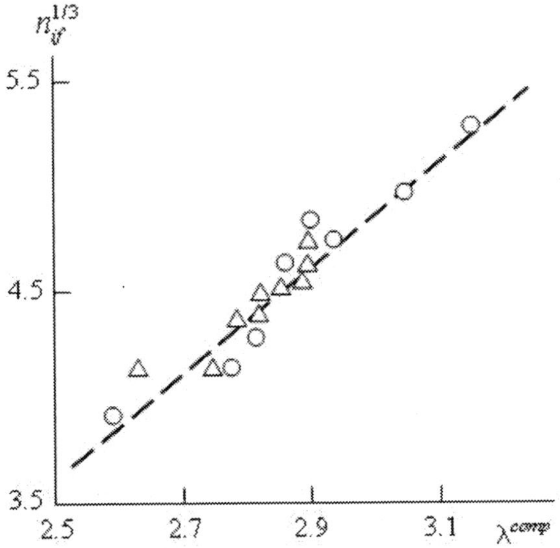

Figure 1.53. The dependence of macromolecular coils number n_{if} in interfacial layer on their compression coefficient λ^{comp}. The designations are the same as in Figure 1.50 [96].

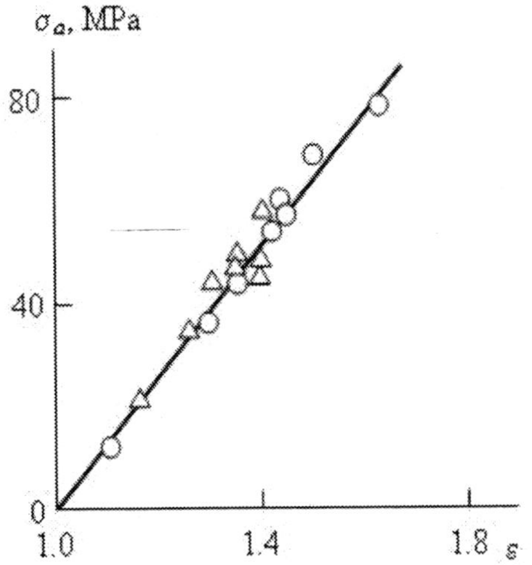

Figure 1.54. The dependence of interfacial layer strength σ_a on exponent ε. The designations are the same as in Figure 1.50 [96].

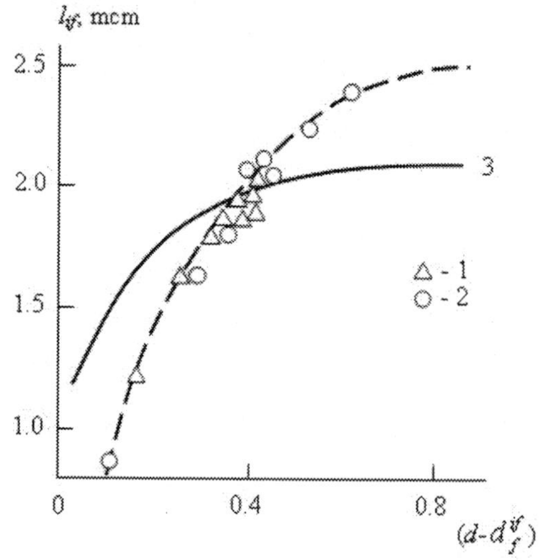

Figure 1.55. The dependences of interfacial layer thickness l_{if} on difference (d- d_f^{if}) for carbon plastics on the basis of phenylone, prepared with magnetic (1) and mechanical (2) separation application. 1, 2 – the experimental data, 3 – calculation according to the relationship (1.49) [100].

As it has been noted above, interfacial layer strength σ_a defines to a considerable extent polymer composites mechanical properties (see the equation (1.42)). In Figure 1.54 the dependence σ_a, calculated according to the equation (1.7), on exponent ε is shown, from which σ_a linear growth at ε increase follows. Therefore, intensification of diffusive processes at interfacial layers formation contributes to their strength rise. The physical ground of this effect is considered above – diffusion intensification contributes to macromolecular coils interpenetration, denser physical entanglements network formation that increases interfacial layer strength [61]. From the said above it follows, that macromolecular coil compactness decreasing (D_c reduction) results to coil compression greatest degree λ^{comp} growth according to the equation (1.67), coils number in interfacial layer rising according to the equation (1.68) and diffusive processes intensification. Let's note that polymerization degree N increasing according to the equation (1.67) gives the similar result. Besides, from the equations (1.4) and (1.30) it follows that molecular characteristics S and C_∞ decrease will also give D_c reduction, that will result to λ^{comp} growth [96].

Therefore, the offered approach demonstrated fractional derivatives theory and fractal analysis successful application for the description of formation of interfacial layers in carbon plastics on both molecular and macroscopic levels, closely connected with one another. Macromolecular coil compactness reduction, characterized by its fractal dimension decreasing, and molecular parameters (macromolecule cross-sectional area and characteristic ratio) decreasing also resulted to interfacial layer thickness growth. Diffusive processes, defined their strength play a very important role in the indicated layers formation.

Let's consider in the present section conclusion interfacial layers structure comparative analysis for polymer composites, filled with short fibers and disperse particles, with fractal analysis notions using [100]. Let's consider the influence of dimensions difference (d- d_f^{if}), where d=3, on interfacial layer thickness l_{if} in the studied composites (carbon plastics on the basis of phenylone and PHE-Gr). In Figure 1.55 the dependence of l_{if} on parameter (d- d_f^{if}), where d=3, for carbon plastics on the basis of phenylone is shown. As one can see, l_{if} increase at (d- d_f^{if}) growth is observed, which tries to attain the layer limited thickness $l_{if}^{\lim}$ at enough large values (d- d_f^{if}). The latter can be estimated from the equation (1.48), accepting limited value φ_{if}=0.74 [8]. In this case $l_{if}^{\lim}$ =2.62 mcm.

For composites PHE-Gr l_{if} values are tabulated in paper [26] and d_f^{if} values can be estimated as follows. Since graphite particles aggregates at large enough φ_f have large values d_{surf} of order 2.70 [24, 39], then according to paper [31] it is supposed that macromolecule in interfacial layer preserves statistical coil conformation with dimension D_c=1.92 [101]. Then the greatest fractal dimension $d_{f\,\max}^{if}$ on interfacial boundary polymer-filler can be determined according to the equation (1.30):

$$d_{f\,\max}^{if} = 1.5D_c = 2.88. \quad (1.69)$$

Further, assuming linear gradient of interfacial layer fractal dimension over its thickness [101], let's determine the average value d_f^{if} for composites PHE-Gr [101]:

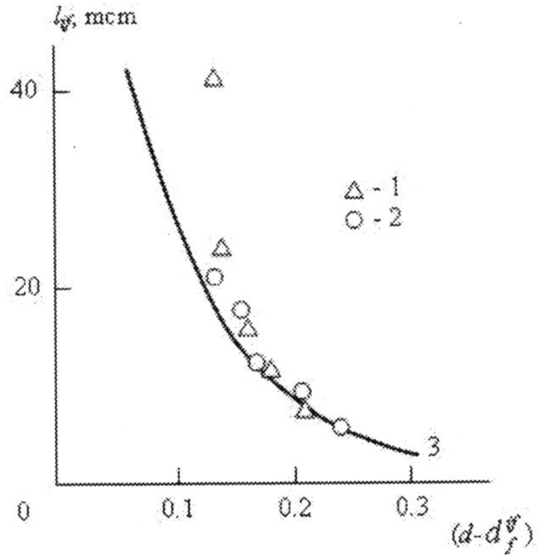

Figure 1.56. The dependences of interfacial layer thickness l_{if} on difference (d- d_f^{if}) for composites PHE-Gr-I (1) and PHE-Gr-II (2). 1, 2 – the experimental data, 3 – calculation according to the relationship (1.49) [100].

$$d_f^{if} = \frac{d_f + d_{f\,\max}^{if}}{2}, \quad (1.70)$$

where d_f is fractal dimension of bulk polymeric matrix, the values of which are adduced in paper [41].

In Figure 1.56 the dependence of l_{if} on (d- d_f^{if}) for composites PHE-Gr is adduced, where tendency, which is opposite to the shown in Figure 1.55 for carbon plastics, is observed, namely, fast reduction of interfacial layer thickness at (d- d_f^{if}) increasing. To explain this dependences l_{if}(d- d_f^{if}) distinction for the considered composites, the authors [100] gave the value l_{if} theoretical estimation within the frameworks of fractal analysis. The theoretical curves l_{if}(d- d_f^{if}), obtained by the best cuperposition of experimental data and calculation according to the relationship (1.49) method, are shown in Figs. 1.55 and 1.56. As one can see, they correspond completely to the experimental results tendency. The discrepancy of theoretical and experimental values l_{if} (not exceeding 40 %) for carbon plastics is due to the fact, probably, that at this parameter calculation fiber radius weak change owing to possible reduction to fragments in preparation process [25, 28] was not accounted for. The relationship (1.49) allows to determine structural grounds of different dependences l_{if}(d- d_f^{if}) course for composites, filled with short fibers and disperse particles. In the first case fiber relative smooth surface and some considerable absence, of filler aggregation due to preliminary blending technology result to the fact, that C_∞ variation (or l_{st}), the value of which grows in virtue of C_∞ increase at d_f^{if} decreasing or (d- d_f^{if}) raising, is the only factor, influencing on the value l_{if}. For composites PHE-Gr at difference (d- d_f^{if}) decreasing the simultaneous C_∞ decreasing, i.e. coil compactness degree in interfacial layer reduction [19], particles aggregates radius R_{ag} and d_{surf}, that is the consequence of graphite particles aggregation [73]. To estimate the distance L_p between filler particles (particles aggregates) surfaces for composites PHE-Gr can be possible according to the equation [65]:

$$L_p = 2R_{ag}\left[\left(\frac{\eta_{dp}}{\varphi_f}\right)^{1/3} - 1\right], \tag{1.71}$$

where η_{dp} is lattice packing density, which is accepted equal to 0.74 [65].

At R_{ag}=40 mcm and φ_f=0.176 L_p value is equal to ~ 49 mcm, i.e. the condition $2l_{if}$ (~ 80 mcm)>L_p (~ 49 mcm) is fulfilled. In other words, in this case interfacial layer widening is observed and polymeric matrix structure becomes identical to interfacial layer structure. For carbon plastics the opposite effect is observed: at (d- d_f^{if})$\to$0 the value $l_{if}\to$0, i.e. interfacial layers collapse is realized and their structure is quite undistinguished from phenylone bulk matrix structure. This factor exercises decisive influence on the interfacial layers strength in the studied composites: if for carbon plastics σ_a grows at l_{if} increasing (see Figure 1.32), then for composites PHE-Gr the opposite effect is observed [26].

Nevertheless, the interfacial layers strength σ_a for both considered composites classes changes equally at the difference (d- d_f^{if}) variation, namely, the σ_a growth at (d- d_f^{if}) increasing is observed (Figure 1.57). As it was noted above, (d- d_f^{if}) growth means interfacial layer energetic component increase [54], that results to σ_a rising. Let's note, that energetic component increasing shows well as in surface energy raising, which in virtue of this, as adhesional characteristics [75], is a structurally dependent parameter.

Therefore, for polymer composites interfacial layers the fractal dimension of substance distribution in them d_f^{if} (or difference (d- d_f^{if})) is the factor, defining these layers strength and, hence, the entire composite strength. At the same time concrete interconnection of this dimension with interfacial layer characteristics (for example, its thickness) is defined by structural parameters of filler and matrix polymer and composites preparation technology also [100] too.

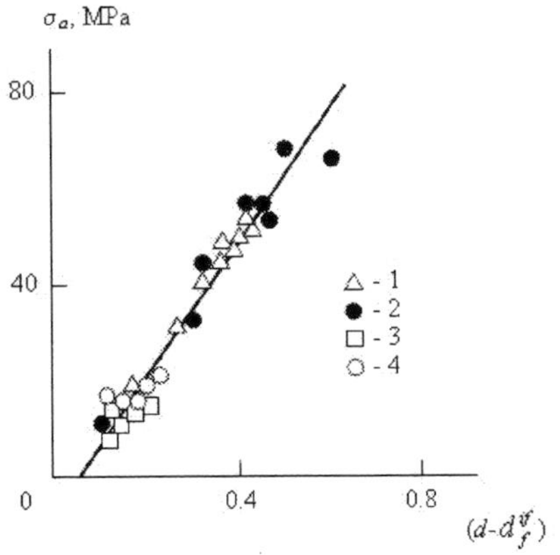

Figure 1.57. The dependence of interfacial layer strength σ_a on difference (d- d_f^{if}) for carbon plastics on the basis of phenylone, prepared with magnetic (1) and mechanical (2) separation application, composites PHE-Gr-I (3) and PHE-Gr-II (4) [100].

1.3. The Percolation Theory Application for Composites Structure Description

As it was shown in papers [16, 102], the amorphous polymers structure, described by the cluster model of polymers amorphous state structure [3, 4], can be presented as percolation system, the critical temperature of which T_{cr} is glass transition temperature T_g. In this case a relative fraction of local order domains (clusters) φ_{cl} is connected with T_g value by the relationship [16]:

$$\varphi_{cl} \approx 0.03(T_g - T)^{0.55}, \qquad (1.72)$$

where T is the testing temperature.

The obtained in papers [16, 102] critical indices β, though they are close to classical percolation indices of order parameter β_{or} (which are equal to 0.37-0.40 [103]), nevertheless are distinguished from them by absolute value. This discrepancy is due to to the fact that percolation cluster presents purely geometrical construction, i.e. it is a too simplified model for the real amorphous polymers, possessing thermodynamically nonequilibrium structure.

Therefore the indicated polymers structure is influenced by thermal interactions (let's remind that cluster structure is postulated as thermofluctuational system [3, 4]). Besides, clusters formation is studied not on the concentrations scale, as for percolation cluster [103], but on the relative temperatures scale. Hence, to simulate amorphous polymers structure as thermal cluster more correctly, i.e. as cluster, whose equilibrium configuration is defined by both geometrical and thermal interactions [18]. In polymeric materials a molecular mobility should be understood as thermal interactions, i.e. thermal oscillations of macromolecules fragments around their quasiequilibrium positions. In this case polymer structure φ_{cl} value is described by the relationship (1.3), in which index β_T is not necessarily equal to the corresponding critical index β_p in purely geometrical percolation models.

Proceeding from this, the authors [104, 105] estimated absolute values β_T variation, studied this index possible connection with polymeric matrix structural characteristics and determined factors, influencing on β_T value in carbon plastics on the basis of phenylone and also compared the obtained results with the data of paper [12] for particulate-filled polymer composites.

In paper [65] it has been shown that critical indices universality for percolation system is connected with its fractal dimension. Percolation system self-similarity supposes subsets availability, having the order n (n=1, 2, 4,...), which in case of polymer composites structure are identified as follows [12, 42]. The first subset (n=1) in polymeric matrix is percolation cluster network or cluster network of matrix physical entanglements; the second (n=2) – loosely-packed matrix, in which cluster network is immersed; the third (n=4) – filler particles network, that is specifical for polymer composites. At such treatment percolation cluster critical indices β_p, ν_p and t_p are given (in three-dimensional Euclidean space) as follows [65]:

$$\beta_n = \frac{1}{d_f}, \tag{1.73}$$

$$\nu_p = \frac{2}{d_f}, \tag{1.74}$$

$$t_p = \frac{4}{d_f}, \tag{1.75}$$

where d_f is a fractal dimension of polymeric matrix structure.

Therefore, the values β_p, ν_p and t_p are border magnitudes for β_T, indicating which composite structural component defines its behaviour. At β_T=β_p such component clusters are or, more precisely, percolation system network identified with cluster network. At $\beta_p<\beta_T<\nu_p$ composite behaviour is due to the combined influence of clusters and loosely-packed matrix. At β_T=ν_p the defining structural component will be loosely-packed matrix, at β_T=t_p – filler particles network and at $\nu_p<\beta_T<t_p$ the combined influence of the two last structural components is observed. Let's note that influence of interfacial layers polymer-filler totality implies as filler particles network influence.

Assuming the average value d_f for the considered composites as equal to ~ 2.50 (see Figure 1.1), from the equations (1.73)-(1.75) well obtain β_p=0.40, ν_p=0.80 and t_p=1.60. Further the value φ_{cl} can be estimated with the aid of the equation (1.4) and, using the calculated by the indicated mode φ_{cl} values, to determine the values β_T for both carbon plastics series according to the equation (1.3).

The dependence of the value β_T on duration t of composite components blending in rotating electromagnetic field is shown in Figure 1.58. From the data of this Figure typically synergetic behaviour β_T as a function t follows: at first the periodic behaviour, close to the sine, and with period doubling is observed and then the transition to chaotic behaviour is realized. The attention is attracted by the fact that all calculated β_T values are close to percolation cluster critical indices, i.e. unlike particulate-filled composites PHE-Gr behaviour β_T change for carbon plastics occurs practically discretely. At small t (~ 10 s) the main factor, influencing on composite properties, is macromolecular entanglements cluster network and at t=60 s – filler particles network (interfacial regions). Then at t≥120 s composite behaviour is controlled by loosely-packed matrix properties (chaotic behaviour).

It was shown earlier [12] that β_T value was a function of molecular mobility degree, characterized by fractal dimension of chain part between entanglements D_{ch}. At D_{ch}=1 the indicated part is stretched completely between entanglements nodes, its molecular mobility is inhibited and β_T=β_p. For the greatest mobility D_{ch}=2. The value D_{ch} calculation can be fulfilled with the aid of the equation (1.55). The dependence $\beta_T(D_{ch})$ for carbon plastics on the basis of phenylone is adduced in Figure 1.59. As it was expected [12], β_T linear increasing at dimension D_{ch} growth is observed. However, unlike nonfilled polymers case [4], D_{ch} limited values are not reached at β_T=β_p. At the last condition fulfillment, i.e. in the percolation cluster case, the chains possess certain mobility degree (D_{ch}≈1.18). This gives the possibility to suppose that filler loosens polymeric matrix (probably, on interfacial boundary) even at the greatest local packing density [70]. Therefore, the main distinction between the initial polymer and the corresponding composite matrix consists in the fact, that the value D_{ch} possible variation for matrix is smaller (D_{ch}=1.18-1.75) than for the initial polymer (D_{ch}=1.0-2.0), that is due to filler influence on polymeric, atrix structure [70].

As it was shown above, the specific feature of the studied carbon plastics structure is the availability in them of densely-packed regions of two types: clusters and interfacial regions having relative fractions φ_{cl} and φ_{if}, accordingly. In addition the approximate identity (1.10) is fulfilled. Let's note, that the value 0.74 in the indicated identity corresponds to densely-packed regions fraction, determined for carbon plastics according to the equation (1.3) at the condition β_T=β_p=0.4. This supposes that φ_{if} increasing decreases φ_{cl}, that should be resulted to polymeric matrix structure loosening and molecular mobility intensification in it, i.e. D_{ch} increase. Such supposition is confirmed by the data of Figure 1.60, where the dependence $D_{ch}(\varphi_{if})$ is adduced. From the plot of this Figure it follows, that between D_{ch} and φ_{if} the linear correlation is obtained though with large scatter. At φ_{if}=0 D_{ch}=1.0, but since the limited lower value D_{ch} according to the plot of Figure 1.59 is equal to ~ 1.18, then from the plot $D_{ch}(\varphi_{if})$ the obligatory availability of interfacial layer in the studied carbon plastics with minimum relative fraction φ_{if}=0.11 follows. This explains the fact, that D_{ch} value >1.0 at β_T=β_p (Figure 1.59). At the greatest value φ_{if}=0.74 D_{ch}≈2.0, i.e. in this case clusters in bulk polymeric matrix do not form, that corresponds to the equation (1.10). Nevertheless, as it follows from the equation (1.55) the condition φ_{cl}=0 is unreal, since in this case $C_\infty \to \infty$, that no has physical significance. Hence, one can suppose that in polymeric matrix dynamic local order exists, similar to order in devitrificated polymers [106].

Therefore, the stated above results showed thermal cluster model adequateness for description of polymer composites structure, filled with short fibers. The thermal and percolation clusters critical indices $\beta_T=\beta_p$ equality is realized in composites at nonzero molecular mobility. The important observation is the dependence $\beta_T(D_{ch})$ universality for different composites classes. The variation range of D_{ch} for composite polymeric matrix is smaller than for the initial polymer because of filler influence on its structure. Hence, this example also shows, that on the studied carbon plastics properties one can influence goal-directly by duration variation of components blending in rotating electromagnetic field [105].

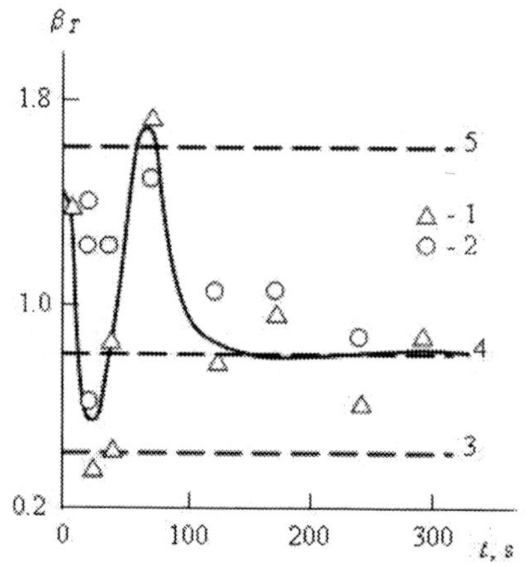

Figure 1.58. The dependence of thermal cluster index β_T on duration t of components blending in rotating electromagnetic field for carbon plastics on the basis of phenylone, prepared with magnetic (1) and mechanical (2) separation application. The horizontal shaded lines are indicated percolation critical indices β_p (3), ν_p (4) and t_p (5) [105].

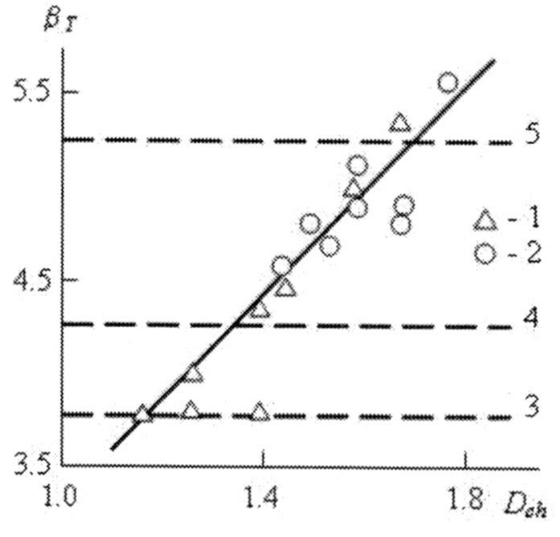

Figure 1.59. The dependence of thermal cluster index β_T on fractal dimension of chain part between entanglements D_{ch} for carbon plastics on the basis of phenylone, prepared with magnetic (1) and mechanical (2) separation application. The horizontal shaded lines indicate percolation critical indices β_p (3), ν_p (4) and t_p (5) [105].

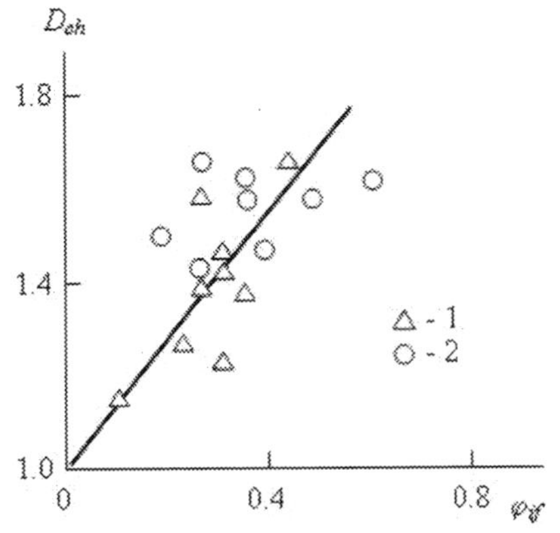

Figure 1.60. The dependence of fractal dimension of chain part between entanglements D_{ch} on interfacial layer relative fraction φ_{if} for carbon plastics on the basis of phenylone, prepared with magnetic (1) and mechanical (2) separation application [105].

The authors of paper [107] used the thermal cluster model for the description of interfacial regions formation in carbon plastics on the basis of phenylone.

In Figure 1.61 the dependence $\varphi_{if}(\beta_T)$ for both series of the studied carbon plastics is shown. Besides, in this Figure by vertical shaded lines percolation critical indices β_p, ν_p and t_p values are indicated. As one can see, the dependence $\varphi_{if}(\beta_T)$ has nonmonotonic character. At small β_T, when cluster network defines composite structure behaviour, the fast growth φ_{if} from its smallest value (~ 0.11) is observed. The interfacial regions fraction reaches its greatest value at $\beta_T \approx 1$, i.e. totality of loosely-packed matrix and filler particles network is defined by structure behaviour. And at last, if filler particles network is defined by composite structure behaviour, then interfacial regions relative fraction decreases fast.

In Figure 1.62 the dependence $\varphi_{l.m.}(\beta_T)$ is adduced, which is an explaining supplement to the plot of Figure 1.61. First of all, comparison of the dependences $\varphi_{if}(\beta_T)$ and $\varphi_{l.m.}(\beta_T)$ shows their clearly expressed antibatness. This supposes that interfacial regions "supply" by polymeric material is realized just at the expence of polymeric matrix loosely-packed regions. From the comparison of the plots of Figs. 1.61 and 1.62 it also follows that the greatest φ_{if} value is reached at the optimal combination of loosely-packed matrix (polymeric material "supplier") and filler particles network (interfacial borders "supplier") joint influences.

As it is known (see Figure 1.58), β_T value is directly proportional to fractal dimension D_{ch} of the chain part between clusters, which characterizes molecular mobility degree in composites [108]. Therefore, from the data of Figs. 1.61 and 1.62 complex enough influence of molecular mobility follows, which is a termal cluster distinctive feature [18], on composites structure formation. At D_{ch} small values molecular mobility contributes to clusters formation, at D_{ch} values, approaching to 2 – to loosely-packed regions formation and at D_{ch} intermediate values interfacial regions are formed. Therefore, the interfacial regions structure by their characteristics should be intermediate between clusters and loosely-packed matrix [107].

As it was shown in paper [98], the relation between interfacial layer mean-square thickness l_{if} and particles number in it n_i was given by the scaling expression (1.64), where exponent ε, characterizing diffusive processes role in interfacial layer formation, was determined according to the equation (1.65). In Figure 1.63 the dependence $\varepsilon(\beta_T)$ for the studied carbon plastics is adduced, which also has nonmonotonic character. It is significant that ε value at any obtained β_T magnitudes does not reach the indicated above limited values (1.0 and 1.70). This means the mixed mechanism of interfacial regions formation, in which diffusive processes play now larger, now smaller role. It is interesting to note, that ε=1.0 is reached at D_{ch} limiting values only: 1.0 and 2.0. In the first from the indicated cases (D_{ch}=1.0) a chain part which is stretched completely between entanglements nodes, loses its molecular mobility and fractal nature [109]. In the second case (D_{ch}=2.0) the chain has the greatest molecular mobility, which is typical for polymeric melt. Since in the considered case polymeric matrix is not oriented then practical significance has the second from the considered cases (D_{ch}=2.0). It has the following physical treatment: in melt possessing by high molecular mobility the chain is stretched on fiber relatively smooth surface, joning with it by physical and/or chemical bonds [31]. This supposes any diffusive processes absence and such mechanism corresponds to criterion ε=1.0. Further on this stretched macromolecule the next one is imposed and so on. At increasing of such normally imposed to fiber surface macromolecules number the interfacial region loosens owing to which diffusive processes begin to play an essential role. This is confirmed by similarity of the curves $\varphi_{if}(\beta_T)$ and $\varepsilon(\beta_T)$ (Figs. 1.61 and 1.63, respectively). At $\beta_T\approx1.0$ a diffusive processes role is quite essential. The subsequent β_T and, hence, D_{ch} reduction makes diffusive processes difficult similarly to temperature decreasing [61] and results to ε decrease.

As it is known [110], in polymeric melt a macromolecule exists in a macromolecular coil form and there are no reasons to assume its complete unfolding in a straight line. Therefore, as above, we suppose that interfacial regions consist of "flattened" under the influence of interaction filler surface coil and then interactions of actually macromolecular coils, having a rotation ellipsoid form. The degree of "flatteness" (compression) of macromolecular coil can be determined as ratio of ellipsoid axes λ^{comp}, which can be calculated according to the equation (1.67), and the value of macromolecular coil in interfacial layer fractal dimension D_c^{if} can be estimated according to the equation (1.30).

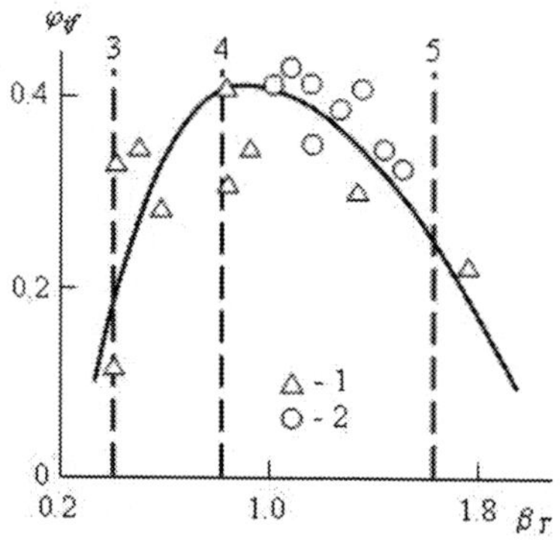

Figure 1.61. The dependence of interfacial regions relative fraction φ_{if} on thermal cluster index β_T value for carbon plastics on the basis of phenylone, prepared with magnetic (1) and mechanical (2) separation application. The vertical shaded lines indicate to percolation critical indices β_p (3), ν_p (4) and t_p (5) [107].

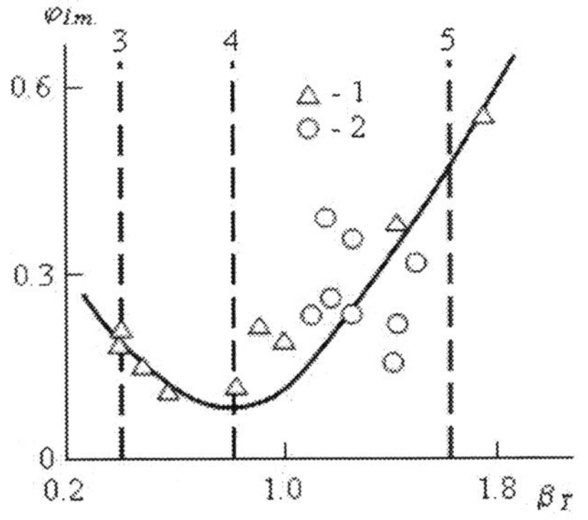

Figure 1.62. The dependence of polymeric matrix loosely-packed regions relative fraction $\varphi_{l.m.}$ on thermal cluster index β_T value for carbon plastics on the basis of phenylone. The designations are the same as in Figure 1.61 [107].

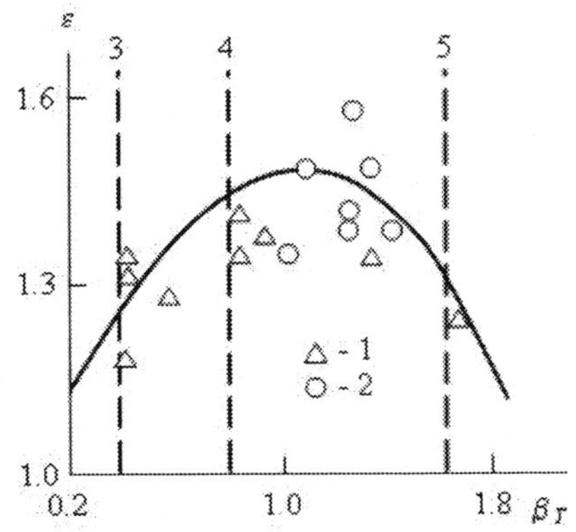

Figure 1.63. The dependence of exponent ε on thermal cluster index β_T value for carbon plastics on the basis of phenylone. The designations are the same as in Figure 1.61 [107].

In Figure 1.64 the dependence $\lambda^{comp}(\beta_T)$ is adduced, which also has nonmonotonic character. Let's note similarity of the dependences $\varepsilon(\beta_T)$ and $\lambda^{comp}(\beta_T)$ (Figs. 1.63 and 1.64, respectively), which shows that diffusive processes also participate in macromolecular coils formation in interfacial regions, having a rotation ellipsoid form. Such ellipsoids formation at $\beta_T \approx 1$ results to the most dense interfacial regions packing, d_f^{if} reduction, l_{if} increasing and, as consequence, to interfacial layer strength growth [107].

Therefore, the stated above results have shown that thermal cluster model and fractal analysis combined application allows to elucidate formation mechanisms, macromolecules conformations and structure specific features for carbon plastics on the basis of phenylone, filled with short carbon fibers.

At present, as a rule, the main structural characteristic of polymer composites is considered their filling volume degree φ_f [70]. However, this parameter is not composites structure universal informant, particularly for composites, filled with short fibers with inherent in them strong anisotropy. The microscopic studies of real composite materials convincingly show fiber uneven distribution, their aggregation, departure from fibers mutual parallelism, porosity availability and so on [111]. Therefore in practice such idea as that or the other property of fiber system, which includes not only physical properties, but geometric features of composite material [111]. Nevertheless, it is obvious, that it is more productive not to operate by such generalized property, but by structural characteristics of fibers system (network), which defines any of its property. Within the frameworks of fractal analysis as such structural characteristic the fractal dimension of filler particles network D_n is used, which describes the density of polymeric matrix space filling with particles or fibers of filler [41, 42]. Carbon plastics studied on the basis of phenylone in the present paper are particularly suitable for the study of the influence of fibers system structure on composites properties as because of technology of their preparation at constant nominal value φ_f they have enough distinguishing properties, for example, their elasticity modulus is changed in the limits 2.13-3.33 GPa. Therefore the authors of paper [112] formulated and determined the effective volume filling degree and described its influence on polymeric matrix properties for the mentioned carbon plastics with the usage of fractal analysis and percolation theory methods.

As it was noted above, the change of components blending duration t in rotating electromagnetic field results to substantial variation of elasticity modulus E_c for the studied carbon plastics (E_c=2.13-3.33 GPa). Within the frameworks of percolation theory the authors [65] proposed the following equation for value E_c determination:

$$\frac{E_c}{E_m} = 1 + 11\varphi_f^{1.7}, \qquad (1.76)$$

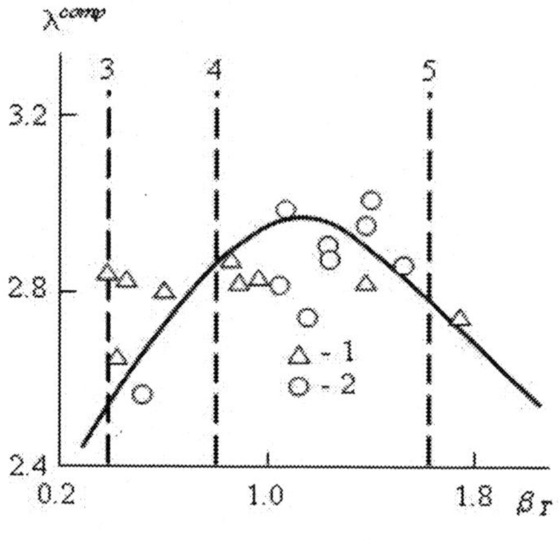

Figure 1.64. The dependence of macromolecular coil compression coefficient λ^{comp} on thermal cluster index β_T value for carbon plastics on the basis of phenylone. The designations are the same as in Figure 1.61 [107].

where E_m is elasticity modulus of matrix polymer.

It is obvious, that the equation (1.76) in its initial form is inapplicable for the description of E_c behaviour for carbon plastics as for them φ_f=const, but E_c value changes in more than 1.5 times. However, this equation can be used for the calculation of the effective volume filling degree φ_f^{ef}, which reflects structure of filler fibers system, defining macroscopic properties of composite (in the considered case E_c). Besides, a the comparison of this parameter with fractal dimension of filler particles network D_n is of interest, which can be determined according to the equation (1.31).

The correlation of structural characteristics of fiber system (network) in carbon plastics on the basis of phenylone φ_f^{ef} and D_n is shown in Figure 1.65. As one can see, the linear growth φ_f^{ef} at the increase D_n is observed. Since D_n is changed within the limits $2.0 \le D_n < 3.0$ [42], then this gives the possibility to estimate variation of magnitudes φ_f^{ef} in the interval 0.060-0.264. The correlation between φ_f^{ef} and D_n is analytically expressed as follows [112]:

$$\varphi_f^{ef} = 0.060 + 0.204(D_n - 2). \quad (1.77)$$

Let's note, that for the considered carbon plastics the inequality $\varphi_f^{ef} > \varphi_f$ (Figure 1.65) is carried out, i.e. the effective value of volume filling degree is always more than nominal one.

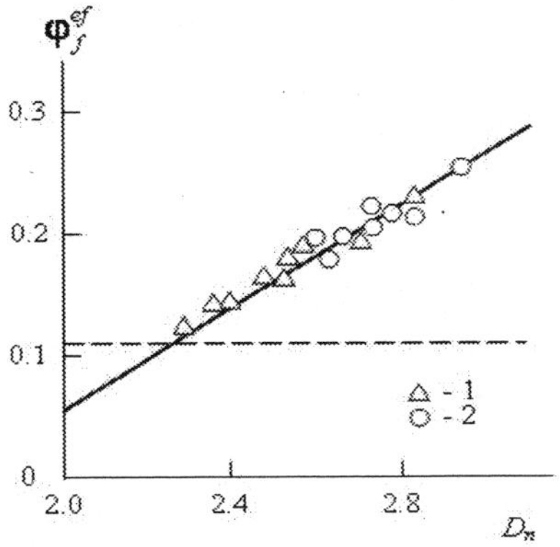

Figure 1.65. The dependence of filling effective volume degree φ_f^{ef} on fibers network fractal dimension D_n for carbon plastics on the basis of phenylone prepared with magnetic (1) and mechanical (2) separation application. The horizontal shaded line points out φ_f nominal value [112].

In Figure 1.66 the dependence φ_f^{ef} on duration t of components blending in rotating electromagnetic field for the considered carbon plastics is adduced. The dependence $\varphi_f^{ef}(t)$ form is specific for synergetic structures: at first periodic (ordered) behaviour φ_f^{ef}, is observed close to sinusoidal one with period doubling and then transition to chaotic behaviour is realized [1]. Let's note, that the value φ_f^{ef} for carbon plastics prepared with mechanical separation application is in average on 20 % more, than the corresponding values for samples prepared with magnetic separation application. This observation supposes an influence of ferromagnetic particles wear products, remaining in composite at the usage of mechanical separation either directly on fibers network structure or (that is more probable) on polymeric matrix structure and then through it – on the mentioned network structure.

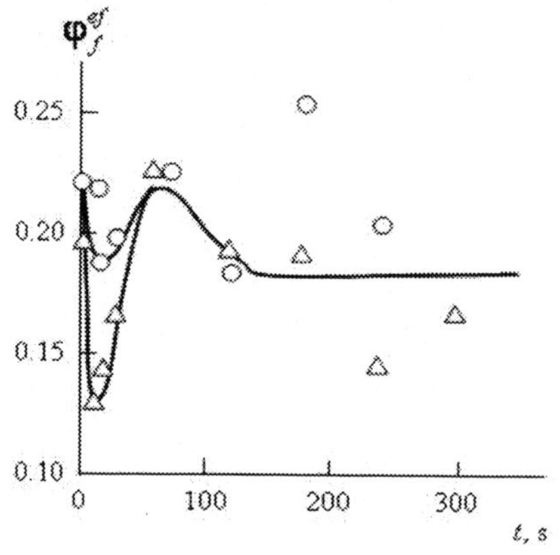

Figure 1.66. The dependence of filling effective volume degree φ_f^{ef} on duration t of components blending in rotating electromagnetic field for carbon plastics on the basis of phenylone. The designations are the same as in Figure 1.65 [112].

As a rule, at present the efficiency of composites filling is described with the usage of modulus efficiency coefficient k_e, which is determined according to the equation [114]:

$$k_e = \frac{E_c - E_m(1-\varphi_f)}{E_f \varphi_f}, \tag{1.78}$$

where E_f is elasticity modulus of filler, which is equal to ~ 15 GPa for carbon fibers [25].

In Figure 1.67 the comparison of two characteristics of carbon plastics filling efficiency k_e and φ_f^{ef} is adduced. As one can see, these parameters are identical and differ only quantitatively: the interconnection between them is given by the simple relationship [112]:

$$k_e = 3.80\varphi_f^{ef}. \tag{1.79}$$

At the same time the principal distinction between parameters k_e and φ_f^{ef} exists, which expresses such distinction of the two existing conceptions of polymer composites reinforcement: classical (continuous mechanics) [78] and fractal conceptions [41]. Within the frameworks of the first conception the dependence E_c on E_f is introcuced that is expressed in the equation (1.78). The fractal conception supposes that the change E_c is due to the modification ("disturbance") of polymer matrix at filler introduction and then the value E_f in this conception does not appear. Let's note the absence E_f in the equation (1.76), obtained within the frameworks of percolation theory. From the fractal conception of filled polymers reinforcement the interconnection of filler particles (fibers) network structure and polymer matrix structure follow. The main informant of substance state is fractal dimension d_f, which in case of polymeric matrix can be determined according to the equations (1.1) and (1.2). In

Figure 1.68 the dependence $d_f(\varphi_f^{ef})$ is adduced, from which d_f growth at φ_f^{ef} increase follows, i.e. raising of effective volume filling degree amplifies the "disturbance" of polymeric matrix structure as in case of particulate-filled polymer composites [41, 42].

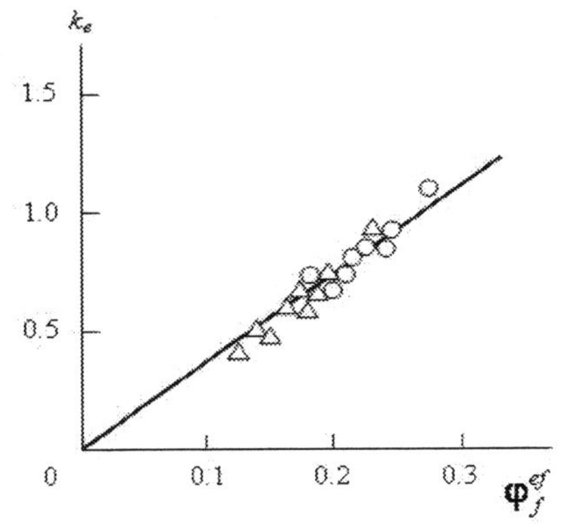

Figure 1.67. The interconnection of modulus efficiency coefficient k_e and effective volume filling degree φ_f^{ef} for carbon plastics on the basis of phenylone. The designations are the same as in Figure 1.65 [112].

Therefore, the present paper results showed, that the structure of short fibers system (network) in carbon plastics could be described by the usage of effective volume filling degree or fractal dimension of this system. An effective filling degree discovers synergetic behaviour and characterizes the filler efficiency for carbon plastics within the frameworks of reinforcement fractal model.

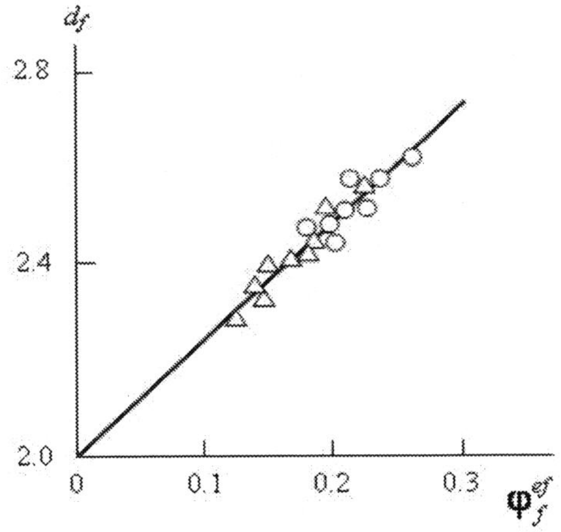

Figure 1.68. The dependence of polymeric matrix structure fractal dimension d_f on the effective volume filling degree φ_f^{ef} for carbon plastics on the basis of phenylone. The designations are the same as in Figure 1.65 [112].

1.4. The Specific Features for Structure of Composites with Semicrystalline Matrix

At present it is supposed that filler introduction in polymer retains structure of the latter invariable and the best variations of structure of crystalline phase [115] or interfacial regions [70] are considered. Such situation is mainly due to the absence until quite recently structural model of polymers amorphous state. However, the appearance of this structure cluster model [3, 4] and fractal analysis application for polymer composites description [24, 39, 41] change the situation in principle. As it is known [116], polymers are structurally complex solids, having several structural levels – molecular, topological, suprasegmental and supramolecular ones. As Academician Kargin very precisely noted "... polymers structure is encoded on molecular level and is realized on supramolecular one". The indicated structure levels are closely interconnected between each other in such a way, that the change of one of them causes the change of the remaining. It is necessary to account for polymer composites structure specific features, one of the main peculiarities of which is interfacial regions polymer-filler availability. Therefore in paper [117] the interconnection was demonstrated and given quantitative description of structural changes on all mentioned levels with the usage of polymers amorphous state structure cluster model and fractal analysis on the example of composites on the basis of semicrystalline high density polyethylene (HDPE) filled with short carbon fibers.

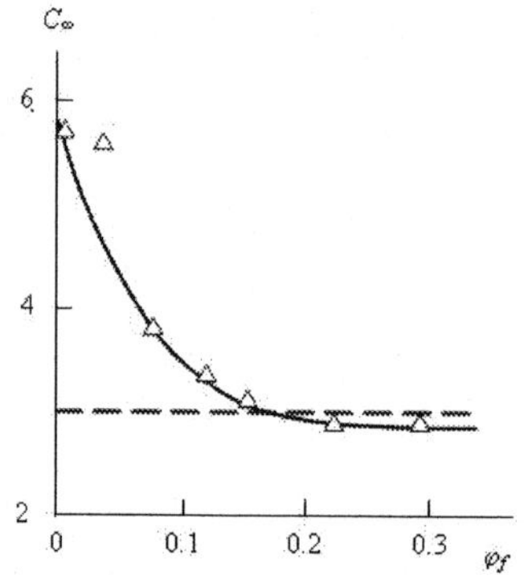

Figure 1.69. The dependence of characteristic ratio C_∞ on filler volume contents φ_f for carbon plastics on the basis of HDPE. The horizontal shaded line points out lower limiting C_∞ value for HDPE [117].

The matrix structural changes consideration of carbon plastics on the basis of HDPE by the mentioned above reasons ought to begin with molecular level. Since polymers present substance consisting of long chain macromolecules then main molecular characteristic ought to consider characteristic ratio C_∞, which is an indicator of polymer chain statistical flexibility [19]. C_∞ connection with polymer structure is confirmed by the fact, that this parameter is a structure automodelity coefficient, i.e. it defines its hierarchy linear scales [2]. C_∞ value can be calculated according to the equation (1.16). In Figure 1.69 the dependence C_∞ on volume degree of filling is adduced, from which C_∞ fast reduction at relatively small φ_f follows and then the dependence $C_\infty(\varphi_f)$ achievement on asymptotic branch at $C_\infty \approx 3$ is observed. The last effect can be easily predicted theoretically. As it is known [3], C_∞ decreasing results to d_f

reduction (the equation (1.16)) and densely-packed regions relative fraction φ_{dens} in amorphous phase increase. However, the last parameter for real polymers never reaches its greatest value 1.0 in virtue of macromolecular nature of polymers. At a certain value φ_{dens} disposing outside densely-packed regions polymer chain sections are stretched and hindered to φ_{dens} subsequent growth. The state, at which the indicated balance is reached, the authors [59] called quasiequilibrium. It is obvious, that this state corresponds to minimum possible d_f value, which is equal to ~ 2.50 for HDPE [59]. From the equation (1.16) it follows that C_∞=3 corresponds to this value, which is the lowest limit of the dependence $C_\infty(\varphi_f)$ (Figure 1.69).

For polymer chain model with fixed valent angles (π-θ) the value C_∞ is determined as follows [19]:

$$C_\infty = \frac{1+\cos\theta}{1-\cos\theta}. \tag{1.80}$$

C_∞ decreasing means θ raising and at C_∞=2 the angle θ becomes tetrahedral [19]. This means that C_∞ reduction is accompanied by polymer chain straightness. Within the frameworks of fractal analysis such effect is supposed at solid smooth surface availability [31], which is filler fibers surface. The overall surface of these fibers will be, obviously, proportional to $\varphi_f^{2/3}$. In Figure 1.70 the dependence θ on parameter $\varphi_f^{2/3}$ is adduced, from which θ growth at relatively small φ_f follows and then reaching of asymptotic branch, is observed which is due to C_∞ similar behaviour (Figure 1.69). It is obvious, that in this case the greatest value θ, allowing by polymer chain chemical constitution, is reached. The dependence $\theta(\varphi_f^{2/3})$ allows also to suppose that for the studied carbon plastics on the basis of HDPE two types of densely-packed regions are possible: interfacial regions and clusters in amorphous phase with relative fractions φ_{if} and φ_{cl}, respectively.

Let's consider further structure topological level changes, which is characterized by traditional ("binary-hooking" or "flings") macromolecular entanglements network density [21]. As it was shown in paper [34], the number of real bonds per chain section between entanglements N_v depends on C_∞ as follows:

$$N_v = 0.36C_\infty^2. \tag{1.81}$$

The equation (1.81) defines N_v decrease at C_∞ reduction and, as consequence, "flings" network density ν_{fl} increasing, which will be proportional to the ratio $(C_\infty^0/C_\infty)^2$, where C_∞^0 is characteristic ratio value for the initial HDPE, which is equal to 5.7 [34]. Accepting ν_{fl}=0.317×10^{27} m^{-3} for the initial HDPE [118], the value ν_{fl} for the studied carbon plastics can be calculated. The dependence $\nu_{fl}(\varphi_f)$ for these materials is shown in Figure 1.71, which demonstrates ν_{fl} growth at φ_f increase. It is necessary to suppose, that the shown in Figure 1.71 φ_{fl} increase is connected with polyethylene chains straightening and, hence, associates with interfacial regions.

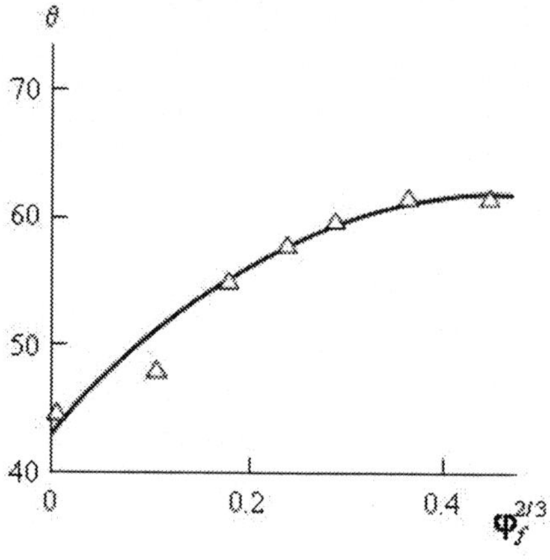

Figure 1.70. The dependence of valent angle θ on parameter $\varphi_f^{2/3}$ for carbon plastics on the basis of HDPE [117].

As it is known [119], ν_{fl} increase acts in crystallization process similarly to polymer molecular weight raising and results to crystallinity degree *K* reduction. The experimental estimation of value *K*, which is crystalline phase integral characteristic, for the studied carbon plastics can be made in supposition of filler ρ_f and polymeric matrix ρ_m densities additivity [117]:

$$\rho_c = \rho_m (1 - \varphi_f) + \rho_f \varphi_f , \qquad (1.82)$$

where ρ_c is composite density.

For carbon fibers ρ_f=1320 kg/m^3 [25]. Further mass crystallinity degree *K* is calculated according to the known formula [120]:

$$K = \frac{\rho_{cr}}{\rho_m} \cdot \frac{\rho_m - \rho_{am}}{\rho_{cr} - \rho_{am}} , \qquad (1.83)$$

where ρ_{cr} and ρ_{am} are polyethylene crystalline and amorphous phases densities, which are equal to 1000 and 850 kg/m^3, respectively.

The theoretical calculation of the value *K* can be fulfilled according to the equation [99]:

In Figure 1.72 the comparison of the obtained experimentally and calculated according to the equation (1.84) *K* values as a function φ_f for the considered carbon plastics is adduced. Both dependence showed *K* reduction at filler content increase C_∞ reduction (Figure 1.70) and ν_{fl} raising (Figure 1.71) that was expected. The data of Figure 1.72 illustrate supermolecular (crystalline) HDPE structure change at filler introduction.

Let's consider further polymeric matrix structure supersegmental level, characterized by parameters φ_{dens} or φ_{if} and φ_{cl}. The value φ_{dens} can be determined with the aid of the equation (1.4). As it is known [121], the crystalline regions in semicrystalline polymers influence essentially on noncrystalline regions structure formation. Within the frameworks of fractal analysis this interconnection is expressed analytically by the following simple relationship [122]:

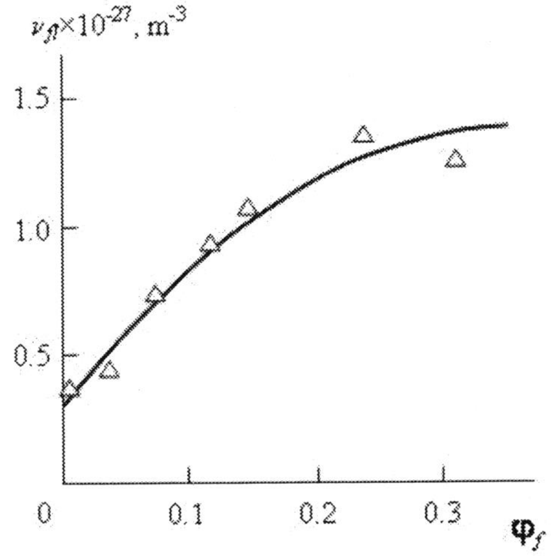

Figure 1.71. The dependence of macromolecular "flings" network density ν_{fl} on filler volume contents φ_f for carbon plastics on the basis of HDPE [117].

$$K = 0.32C_{\infty}^{1/3}. \tag{1.84}$$

$$d_f \approx 2 + K, \tag{1.85}$$

since, as it follows from the equation (1.4), the value d_f is controlled by polyethylene structure supersegmental level.

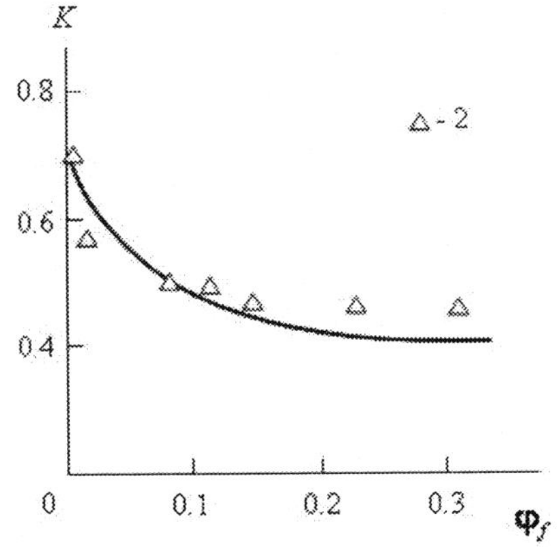

Figure 1.72. The dependences of experimental (1) and calculated according to the equation (1.84) (2) crystallinity degree K on filler volume contents φ_f for carbon plastics on the basis of HDPE [117].

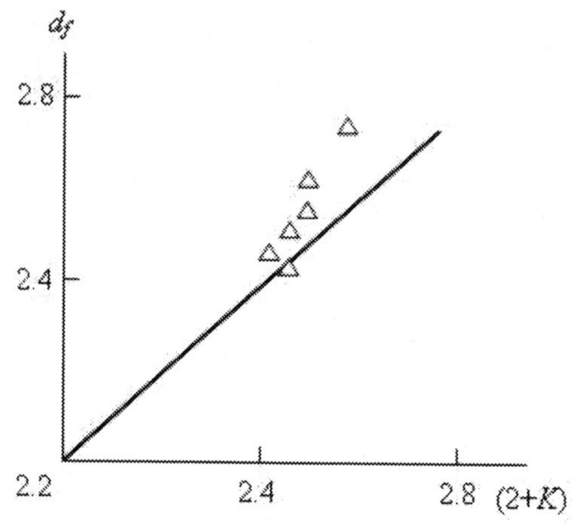

Figure 1.73. The comparison of fractal dimension d_f and parameter (2+K) values for carbon plastics on the basis of HDPE. The straight line shows the relation 1:1 [117].

The adduced in Figure 1.73 comparison of parameters d_f and (2+K) according to the relationship (1.85) actually shows their approximate equality for the studied carbon plastics too. The parameters φ_{cl} and φ_{if} can be estimated separately as follows. As it is known [123], the relation between true stress σ^{tr} on cold flow plateau and drawing ratio λ has the form:

$$\sigma^{tr} = G_p\left(\lambda^2 - \lambda^{-1}\right), \tag{1.86}$$

where G_p is the so-called strain hardening modulus.

Further according to the known G_p values physical entanglements cluster network density ν_{cl} is calculated [4]:

$$\nu_{cl} = \frac{G_p N_A}{RT}, \tag{1.87}$$

where N_A is Avogadro number, R is universal gas constant, T is testing temperature.

And at last, the value φ_{cl} can be determined as follows [4]:

$$\varphi_{cl} = SC_\infty l_0 \nu_{cl}. \tag{1.88}$$

As it follows from the data of Figure 1.74, the value φ_{cl} is practically invariable at φ_f change, whereas the value φ_{dens} and accordingly φ_{if}, determined as (φ_{dens}-φ_{cl}), change similarly: at first their fast growth is observed and then going out on an asymptotic branch. The condition φ_{cl}=const is defined by the laws of local order domains formation in semicrystalline polymers with vitrificated amorphous phase [124]. In this case the value φ_{cl} is proportional to parameter β_{cr}, which characterizes amorphous chains tightness in crystallization process [125]. The parameter β_{cr} is determined according to the relationship [11]:

$$\beta_{cr}^2 = \frac{5EM_{cl}(1-K)^3}{\rho_m RT}, \qquad (1.89)$$

where E is elasticity modulus, M_{cl} is molecular weight of chain part between clusters, which is calculated according to the well-known equation [4]:

$$M_{cl} = \frac{\rho_m N_A}{\nu_{cl}}. \qquad (1.90)$$

As estimations according to the equation (1.89) have shown, the value β_{cr} is approximately constant and changes within the limits of calculation error (in the range ±10 %). In its turn, the condition β_{cr}=constant defines the result φ_{cl}=const.

Therefore, the stated above results have shown that high density polyethylene structure change dynamics at filler (short fibers) introduction consists of its changes on all structural levels. These changes are interconnected and can be described quantitatively within the frameworks of polymers amorphous state structure cluster model and fractal analysis. Let's note, that these changes theoretical description is obtained with the usage of carbon plastics mechanical tests results (see the equations (1.1) and (1.2)). Consequently the opposite is also true: the offered model allows composites properties prediction, if their structural changes are known [117].

In paper [126] the analysis of short carbon fibers influence on HDPE matrix crystallization process was fulfilled. The polymers crystallinity degree change is closely connected with their crystallization kinetics, which can be described by well-known Kolmogorov-Avrami equation [127]:

$$K = 1 - e^{-Zt_{cr}^n}, \qquad (1.91)$$

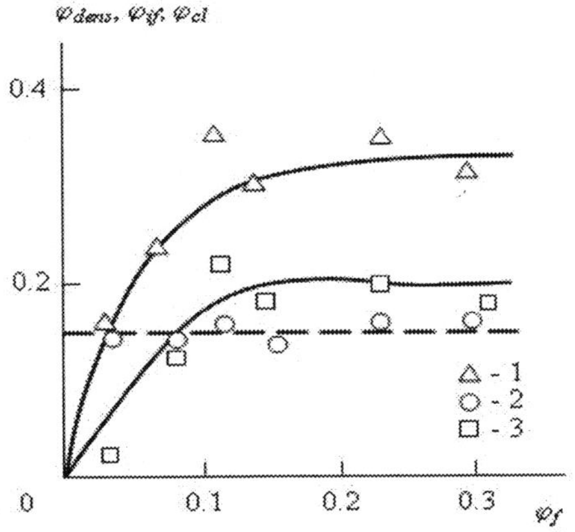

Figure 1.74. The dependences of densely-packed domains φ_{dens} (1), interfacial regions φ_{if} (2) and clusters φ_{cl} (3) relative fractions on filler volume contents φ_f for carbon plastics on the basis of HDPE [117].

where Z is crystallization rate constant, t_{cr} is crystallization process duration, n is Kolmogorov-Avrami exponent, characterizing for the given polymer nucleation and the growing crystalline structures type.

As it was shown in paper [128], the exponent n is connected with fractal dimension of chain part between clusters D_{ch} as follows:

$$n = 3(D_{ch} - 1) + 1 . \qquad (1.92)$$

The value D_{ch} can be calculated according to the equation (1.55) with φ_{cl} replacement on φ_{cl}^{red}, taking into accout filler presence and estimated as follows [126]:

$$\varphi_{cl}^{red} = \frac{\varphi_{cl}}{1 - \varphi_f} . \qquad (1.93)$$

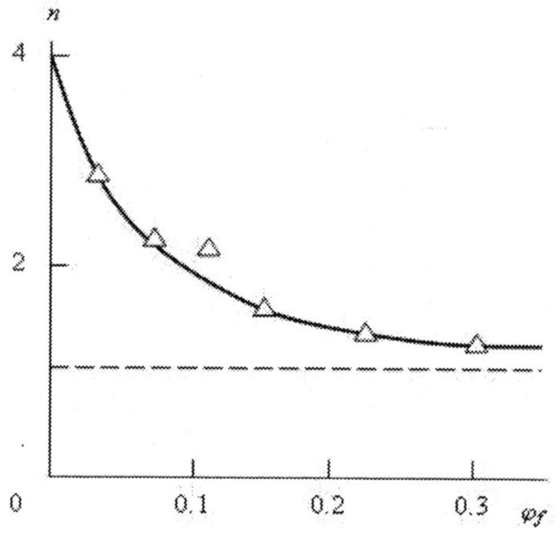

Figure 1.75. The dependence of the exponent n in Kolmogorov-Avrami equation on filler volume contents φ_f for carbon plastics on the basis of HDPE [126].

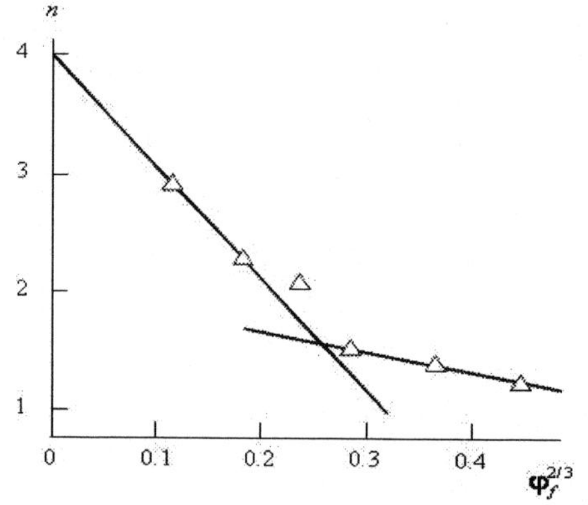

Figure 1.76. The dependence of exponent n in Kolmogorov-Avrami equation on parameter $\varphi_f^{2/3}$, characterizing carbon fibers surface total area, for carbon plastics on the basis of HDPE [126].

In its turn, the value φ_{cl} is calculated according to the equation (1.4).

In Figure 1.75 the dependence of the exponent n, calculated according to the equation (1.92), on φ_f value, which shows n reduction at carbon fibers volume contents growth. As it is known [127], variation n defines morphology of polymers forming crystalline phase. In case of athermic nucleation at $n \le 2.0$ ribbons are formed by two-dimensional growth mechanism, at $n \le 3.0$ - circles and at $n > 3.0$ – spheres. Fractional values n mean a combined mechanism of thermal/athermic nucleation, the fractional part decrease indicates on athermic mechanism role increasing at that.

It is obvious, that in the general case introduction of filler with smooth surface results to crystals growth change from three-dimensional to two-dimensional one with corresponding reduction of the exponent n [127]. This law is easily traced by the following simple mode. The total area of carbon fibers surface will grow proportionaly to $\varphi_f^{2/3}$ and the dependence $n($ $\varphi_f^{2/3})$ adduced in Figure 1.76 shows actually linear reduction n at filler surface total area growth. Probably, at $\varphi_f > 0.25$ fiber aggregation process begins, that changes surface total area and slows down n reduction. At $\varphi_f \approx 0.55$ the value n decreases up to 1.0, i.e. at large φ_f athermic nucleation is realized, in other words, all crystallites simultaneous beginning of growth, evidently, on fibers surface [127]. At $\varphi_f = 0$, i.e. for the initial HDPE, $n = 4.0$ and in this case spherical crystalline forms growth occurs at thermal nucleation, i.e. new crystallites origination in crystallization process [127]. The similar effect was observed for the system polyamide-6/quartz powder, where crystalline phase parameters were linearly reduced at filler specific surface growth [129].

Further the value K was calculated according to the equation (1.91) at arbitrary values $Z=0.145$ and $t_{cr}=2$. As it is shown in Figure 1.77, such calculation gave a good correspondence to the experiment. Let's note, that the condition $Z=0.145=\text{const}$ for all used φ_f assumes crystallization constant rate. In this case the value K at $t_{cr}=\text{constant}$ is defined only by nucleation and growing crystalline structures type, i.e. by the value n, as it was described above.

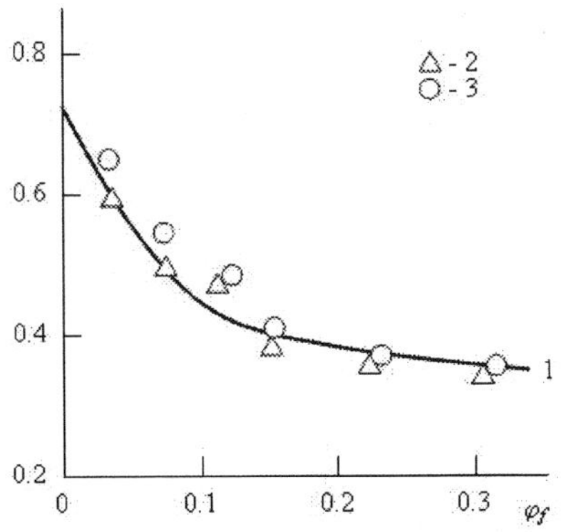

Figure 1.77. The dependence of crystallinity degree K on filler volume contents φ_f for carbon plastics on the basis of HDPE. 1 – the experimental data; 2 – calculation according to the equation (1.91) [126].

As the estimations according to the equations (1.4) and (1.93) have shown, the clusters reduced relative fraction φ_{cl}^{red} increases from 0.119 up to 0.302 at the corresponding reduction K from 0.590 up to 0.330 (Figure 1.72). Therefore, long-range order degree decreasing is compensated by local (short-range) degree growth. In Figure 1.78 the dependence (K+ φ_{cl}^{red}) on the value φ_f is adduced, which assumes the condition (K+ φ_{cl}^{red})=const. As it is known [18], the greatest fraction of densely-packed regions, i.e. semicrystalline matrix (K+ φ_{cl}^{red}), can be calculated within the frameworks of thermal cluster conception by analogy with the equation (1.3) [126]:

$$\left(K+\varphi_{cl}^{red}\right)=\left(\frac{T_m-T}{T_m}\right)^{0.37}, \tag{1.94}$$

where T_m is melting temperature and β_T=β_p=0.37 [103]. According to the equation (1.94) (K+ φ_{cl}^{red})=0.63, that is close to the value 0.66 according to the data of Figure 1.78.

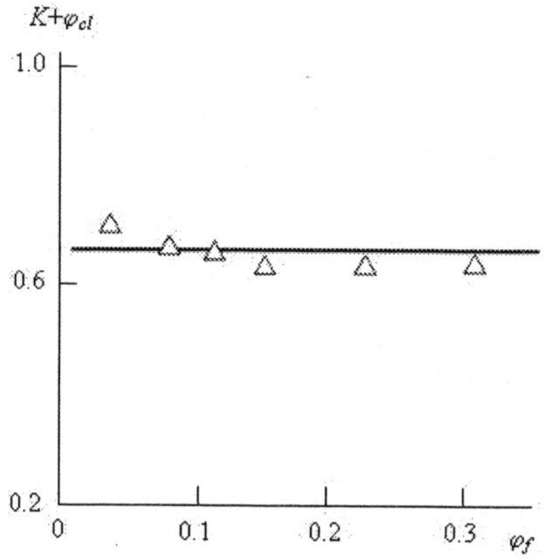

Figure 1.78. The dependence of densely-packed regions total fraction (K+ φ_{cl}^{red}) on carbon fibers volume contents φ_f for carbon plastics on the basis of HDPE [126].

In paper [130] the conditions, at which cluster becomes nucleator have been considered and it was shown that this occurs only, when statistical segment length l_{st} reaches a certain critical value. It is obvious, that calculated according to the equation (1.16) values C_∞ decrease in the increase range φ_f from 0.038 up to 0.303 (Figure 1.69) results to clusters number reduction, capable to become crystallization nucleators in the thermal mechanism case. This factor should also influence on K reduction at φ_f growth.

Consequently, the stated above results illustrate the change dynamics of semicrystalline matrix structure of carbon plastics on the basis of HDPE at carbon fibers introduction. These changes affect both crystalline phase and amorphous one and are closely interconnected between themselves [117, 126].

References

[1] Ivanova V.S., Kuzeev I.R., Zakirnichnaya M.M. *Synergetics and Fractals. Universality of Materials Mechanical Behviour.* Ufa, Publishers USNTU, 1998, 366 p.

[2] Kozlov G.V., Novikov V.U. *Synergetics and Fractal Analysis of Cross-linked Polymers.* Moscow, Klassika, 1998, 112 p.

[3] Kozlov G.V., Novikov V.U. *Uspekhi Fizicheskikh Nauk,* 2001, v. 171, № 7, p. 717-764.

[4] Kozlov G.V., Zaikov G.E. Structure of the Polymer Amorphous State. Utrecht-Boston, *Brill Academic Publishers*, 2004, 465 p.

[5] Burya A.I., Kozlov G.V. Mater. Of Sci.-Pract. Conf. "Science, Engineering and Higher Education". Issue 1. "Development Problems and Tendencies". Rostov-on-Don, *RSU,* 2004, p. 56-58.

[6] Burya A.I., Kozlov G.V., Kazakov M.E. *Mater. of 24th Annual International Conf. "Composite Materials in Industry (Slavpolikom)",* Yalta, 31 May-4 June 2004, p. 246-248.

[7] Burya A.I., Kozlov G.V., Chaika L.V. *Scientific Works of Donetskogo NTU, Khimiya i Khimicheskaya Technologiya,* 2004, № 77, p. 44-50.

[8] Burya A.I., Kozlov G.V. *Voprosy Khimii i Khimicheskoy Technologii,* 2005, № 3, p. 106-112.

[9] Fomichev A.I., Burya A.I., Gubenkov M.G. *Electronnaya Obrabotka Materialov,* 1978, № 4, p. 26-27.

[10] Balankin A.S. *Synergetics of Deformable Body.* Moscow, Publishers Ministry of Defence SSSR, 1991, 404 p.

[11] Kozlov G.V., Sanditov D.S. Anharmonic Effects and Physical-Mechanical Properties of Polymers. Novosibirsk, *Nauka,* 1994, 261 p.

[12] Kozlov G.V., Lipatov Yu.S. *Mekhanika Kompositnykh Materialov,* 2003, v. 39, № 1, p. 89-96.

[13] Kozlov G.V., Zaikov G.E. *Materialovedenie,* 2002, № 12, p. 13-17.

[14] Kozlov G.V., Aloev V.Z. Percolation Theory in Polymers Physics-Chemistry. Nal'chik, *Polygraphservis and T,* 2005, 148 p.

[15] Sokolov I.M. *Uspekhi Fizicheskikh Nauk,* 1986, v. 150, № 2, p. 221-256.

[16] Kozlov G.V., Gazaev M.A., Novikov V.U., *Mikitaev A.K. Pis'ma v ZhTF,* 1996, v. 22, № 16, p. 31-38.

[17] Kozlov G.V., Gazaev M.A., Novikov V.U., Mikitaev A.K. *Inzhenerno-Fizicheskiy Zhurnal,* 1998, v. 71, № 2, p. 241-247.

[18] Family F. *J. Stat. Phys.*, 1984, v. 36, № 5/6, p. 881-896.

[19] Budtov V.P. *Physical Chemistry of Polymer Solutions.* Sankt-Peterburg, Khimiya, 1992, 384 p.

[20] Aharoni S.M. *Macromolecules,* 1985, v. 18, № 12, p. 2624-2630.

[21] Aharoni S.M. *Macromolecules,* 1983, v. 16, № 9, p. 1722-1728.

[22] Tovmasyan Yu.M., Topolkaraev V.A., Berlin Al.Al. *Vysokomolek. Soed.* A, 1986, v. 28, № 6, p. 1162-1167.

[23] Morosova N.V., Topolkaraev V.A. *Vysokomolek. Soed.* A, 1991, v. 33, № 1, p. 81-86.

[24] Novikov V.U., Kozlov G.V., Bur'yan O.Yu. *Mekhanika Kompositnykh Materialov,* 2000, v. 36, № 1, p. 3-32.

[25] Burya A.I., Chigvintseva O.P., Suchilina-Sokolenko S.P. Polyarylates. Synthesis, Properties, Composite Materials. Dnepropetrovsk, *Nauka i Obrazovanie*, 2001, 152 p.
[26] Novikov V.U., Kozlov G.V., Lipatov Yu.S. *Plast. Massy,* 2003, № 10, p. 4-8.
[27] Novikov V.U., Kozlov G.V. *Uspekhi Khimii,* 2000, v. 69, № 6, p. 572-599.
[28] Klimontovich Yu.L. *Turbulent Motion and Chaos Structure: the New Approach to Statistical Theory of Open Systems.* Moscow, Nauka, 1990, 320 p.
[29] Petruchenko A.A. *Freedback Principle.* Moscow, Mir, 1987, 224 p.
[30] Lipatov Yu.S. *Physical Chemistry of Filled Polymers.* Moscow, Khimiya, 1977, 304 p.
[31] Pfeifer P. In book: *Fractals in Physics.* Ed. Pietronero L., Tosatti E. Amsterdam, Oxford, New York, Tokyo, North-Holland, 1986, p. 72-81.
[32] Burya A.I., Kozlov G.V., Chukalovsky P.A., Rula I.V. Mater. 5th International Conf. “Research and Development in Mechanical Industry”. 4-7 Sept. 2005, Vrnjačka Banja, *Serbia and Montenegro,* p. 326-330.
[33] Bovenko V.N., Startsev V.M. *Vysokomolek. Soed. B,* 1994, v. 36, № 6, p. 1004-1008.
[34] Wu S. *J. Polymer Sci.: Part B: Polymer Phys.,* 1989, v. 27, № 4, p. 723-741.
[35] Burya A.I., Kozlov G.V. *Trenie i iznos,* 2003, v. 24, № 3, p. 279-283.
[36] McCayley J.L. *Int. J. Modern Phys. B,* 1989, v. 3, № 6, p. 821-852.
[37] Kozlov G.V., Burya A.I., Zaikov G.E. *J Appl. Polymer Sci.,* 2006, v. 100, № 4, p. 2817-2820.
[38] Kozlov G.V., Burya A.I., Zaikov G.E. In book: Chemical Reactions in Condensed Phase. Quantitative Level. Ed. Zaikov G., Zaikov V., Mikitaev A. New York, *Nova Science Publishera,* Inc., 2006, p. 207-214.
[39] Kozlov G.V., Yanovsky Yu.G., Lipatov Yu.S. *Mekhanika Kompozitsionnykh Materialov i Konstruktsii,* 2008, v. 8, № 1, p. 111-149.
[40] Kozlov G.V., Burya A.I. *Doklady NAN Ukraine,* 2006, № 3, p. 136-142.
[41] Novikov V.U., Kozlov G.V. *Mekhanika Kompositnykh Materialov,* 1999, v. 35, № 3, p. 269-290.
[42] Kozlov G.V., Mikitaev A.K. *Mekhanika Kompozitsionnykh Materialov i Konstruktsii,* 1996, v. 2, № 3-6, p. 144-157.
[43] Kozlov G.V., Yanovsky Yu.G., Mikitaev A.K. *Mekhanika Kompositnykh Materialov,* 1998, v. 34, № 4, p. 539-544.
[44] Avnir D., Farin D., Pfeifer P. *Nature;* 1984, v. 308, № 5959, p. 261-263.
[45] Burya A.I., Kozlov G.V., Kholodilov O.V. Vestnik Polotskogo Gosuniversiteta, *Seriya B,* 2005, № 6, p. 36-39.
[46] Kozlov G.V., Shustov G.B., Burya A.I. *Plast. Massy,* 2006, № 1, p. 11-13.
[47] Burya A.I., Kozlov G.V., Rula I.V. *Novosti Nauki Pridneprov'ya,* 2004, № 3, p. 8-11.
[48] Vannimenus J. *Physica D,* 1989, v. 38, № 2, p. 351-355.
[49] Kozlov G.V., Yanovsky Yu.G., Lipatov Yu.S. *Mekhanika Kompozitsionnykh Materialov i Konstruktsii,* 2002, v. 8, № 4, p. 467-474.
[50] Kozlov G.V., Lipatov Yu.S. *Mekhanika Kompositnykh Materialov,* 2004, v. 40, № 6, p. 827-834.
[51] Baranov V.G., Frenkel' S.Ya., Brestkin Yu.V. *Doklady AN SSSR,* 1986, v. 290, № 2, p. 369-372.
[52] Kozlov G.V., Temiraev K.B., Shustov G.B., Mashukov N.I. *J. Appl. Polymer Sci.,* 2002, v. 85, № 6, p. 1137-1140.
[53] Kozlov G.V., Yanovsky Yu.G., *Mikitaev A.K. Poverkhnost',* 1999, № 8, p. 43-46.

[54] Kuzeev I.R., Samigullin G.Kh., Kulikov D.V., Zakirnichnaya M.M. Complex Systems in Nature and Engineering. Ufa, *Publishers USNTU,* 1997, 225 p.

[55] Kozlov G.V., Burya A.I., Aloev V.Z. Zavodskaya Laboratoriya. *Diagnostika Materialov,* 2005, v. 71, № 11, p. 33-35.

[56] Kalinchev E.L., Sakovtseva M.E. Properties and Processing of Thermoplastics. Leningrad, *Khimiya,* 1983, 288 p.

[57] Burya A.I., Kozlov G.V., Sverdlikovskaya O.S. *Voprosy Khimii i Khimicheskoi Technologii,* 2004, № 4, p. 109-112.

[58] Burya A.I., Chigvintseva O.P. *Sovremennoe Mashinostroenie,* 1999, № 2, p. 28-32.

[59] Kozlov G.V., Zaikov G.E. In book: Fractals and Local Order in Polymeric Materials. Ed. Kozlov G., Zaikov G. New York, *Nova Science Publishers,* Inc., 2001, p. 55-63.

[60] Leidner J., Woodhams R.T. *J. Appl. Polymer Sci.,* 1994, v. 18, № 8, p. 1639-1654.

[61] Schnell R., Stamm M., Creton C. *Macromolecules,* 1998, v. 31, № 7, p. 2284-2292.

[62] Burya A.I., Shogenov V.N., Kozlov G.V., Kholodilov O.V. Materialy. *Technologii. Instrumenty,* 1999, v. 4, № 2, p. 39-41.

[63] Kozlov G.V., Kolodey V.S., Lipatov Yu.S. *Materialovedenie,* 2002, № 11, p. 34-39.

[64] Burya A.I., Kozlov G.V., Sviridenok A.I. *Doklady NAN Belarusi,* 2001, v. 45, № 3, p. 120-122.

[65] Bobryshev A.N., Kozomazov V.N., Babin L.O., Solomatov V.I. Synergetics of Composite Materials. Lipetsk, *NPO ORIUS,* 1994, 154 p.

[66] Abaev A.M., Kozlov G.V., Shustov G.B., Mikitaev A.K. *Izvestiya KBSC RAN,* 2000, № 1(4), p. 104-107.

[67] Yakubov T.S. *Doklady AN SSSR,* 1988, v. 303, № 2, p. 425-428.

[68] Pfeifer P. *Appl. Surface Sci.,* 1984, v. 18, № 1, p. 146-164.

[69] Burya A.I., Kozlov G.V., Lipatov Yu.S. *Materialy. Technologii. Instrumenty,* 2002, v. 7, № 3, p. 42-44.

[70] Lipatov Yu.S. *Physical-Chemical Grounds of Polymers Filling.* Moscow, Khimiya, 1991, 260 p.

[71] Kozlov G.V., Ovcharenko E.N., Lipatov Yu.S. *Doklady NAN Ukraine,* 1999, № 11, p. 128-132.

[72] Kozlov G.V., *Lipatov Yu.S. Poverkhnost',* 2003, № 8, p. 81-84.

[73] Kozlov G.V., Yanovsky Yu.G., Lipatov Yu.S. *Mekhanika Kompozitsionnykh Materialov i Konstruktsii,* 2003, v. 9, № 3, p. 398-448.

[74] Van Damme H., Levitz P., Bergaya F., Alcover J.F., Gatineau L., Fripiat J.J. *J. Chem. Phys.,* 1986, v. 85, № 1, p. 616-625.

[75] Kozlov G.V., Lipatov Yu.S. *Composite Interfaces,* 2002, v. 9, № 6, p. 509-527.

[76] Burya A.I., Kozlov G.V., Vishnyakov L.R. Proceedings of International Conf. "Modern Materials Studies: Achievements and Problems". *MMS-2005.* 26-30 Sept. 2005, Kiev, p. 105-106.

[77] Hentschel H.G.E., Deutch *J.M. Phys. Rev.* A, 1984, v. 29, № 3, p. 1609-1611.

[78] Ahmed S., Jones F.R. *J. Mater. Sci.,* 1990, v. 25, № 12, p. 4933-4942.

[79] Kozlov G.V., Lipatov Yu.S. Voprosy Khimii i Khimicheskoi Technologii, 2002, № 3, p. 65-67.

[80] Kozlov G.V., Burya A.I., Aloev V.Z., Yanovsky Yu.G. *Fozocheskaya Mezomekhanika,* 2005, v. 8, № 2, p. 35-38.

[81] Kubat J., Rigdahl M., Welander M. J. Appl. Polymer Sci., 1990, v. 39, № 5, p. 1527-1539.
[82] Kozlov G.V., Beloshenko V.A., Varyukhin V.N. *Ukrainskii Fizicheskii Zhurnal,* 1998, v. 43, № 3, p. 322-323.
[83] Shogenov V.N., Kozlov G.V. Fractal Clusters in Polymers Physics-Chemistry. Nal'chik, *Polygraphservice and T,* 2002, 270 p.
[84] Lipatov Yu.S. *Interfacial Phenomena in Polymers.* Kiev, Naukova Dumka, 1980, 260 p.
[85] Kozlov G.V., Burya A.I., Zaikov G.E. In book: Molecular and High Molecular Chemistry: Theory and Practice. Ed. Monakov Yu., Zaikov G. New York, *Nova Science Publishers,* Inc., 2006, p. 147-153.
[86] Kozlov G.V., Temiraev K.B., Shetov R.A., Mikitaev A.K. *Materialovedenie,* 1999, № 2, p. 34-39.
[87] Kozlov G.V., Beloshenko V.A., Kuznetsov E.N., Lipatov Yu.S. *Doklady NAN Ukraine,* 1994, № 12, p. 126-128.
[88] Kozlov G.V., Burya A.I., Shustov G.B. Izvestiya VUZov, Severo-Kavkazsk. *region, estestv. nauki,* 2005, № 3, p. 62-65.
[89] Aloev V.Z., Kozlov G.V. *Fizika i Technika Vysokikh Davlenii,* 2001, v. 11, № 1, p. 40-42.
[90] Kozlov G.V., Burya A.I., Dolbin I.V., Zaikov G.E. *J. Appl. Polymer Sci.,* 2006, v. 101, № 6, p. 4044-4047.
[91] Chukbar K.B. *Zhurnal Eksperimental'noi i Teoreticheskoi Fiziki,* 1995, v. 108, № 5, p. 1875-1884.
[92] Shogenov V.N., Akhkubekov A.A., Akhkubekov R.A. Izvestiya VUZov, Severo-Kavkazsk. region, estestv. *nauki,* 2004, № 1, p. 46-50.
[93] Kozlov G.V., Shustov G.B., Zaikov G.E. *J. Balkan Tribologic. Assoc.,* 2003, v. 9, № 4, p. 467-514.
[94] Vilgis T.A., Haronska P., Benhamou M. *J. Phys. II France*, 1994, v. 4, № 12, p. 2187-2196.
[95] Kozlov G.V., Dolbin I.V., Shogenov V.Kh., Zaikov G.E. *J. Appl. Polymer Sci.,* 2004, v. 91, № 6, p. 3765-3768.
[96] Kozlov G.V., Burya A.I., Aloev V.Z. *Konstruktsii iz Kompozitsionnykh Materialov,* 2008, № 2, p. 71-77.
[97] Meakin P. *Phys. Rev.* A, 1983, v. 27, № 5, p. 2616-2623.
[98] Meakin P. *Phys. Rev.* B, 1984, v. 30, № 8, p. 4207-4214.
[99] Aloev V.Z., Kozlov G.V. Physics of Orientational Phenomena in Polymer Materials. Nal'chik, *Polygraphservice and T,* 2002, 288 p.
[100] Kozlov G.V., Burya A.I., Dolbin I.V. *Materialovedenie,* 2005, № 8, p. 31-35.
[101] Kozlov G.V. *Izvestiya KBSC RAN,* 2003, № 1(9), p. 50-53.
[102] Kozlov G.V., Novikov V.U., Dolbin I.V., Zaikov G.E. Proceedings of International Conf. "Baikalian Readings-II by Processes Simulation in Synergetic Systems". Ulan0Ude-Tomsk, *Publishers TSU,* 2002, p. 212-214.
[103] Shklovsky B.I., Efros A.L. *Uspekhi Fizicheskikh Nauk,* 1975, v. 117, № 3, p. 401-436.
[104] Burya A.I., Aloev V.Z., Kozlov G.V. *Uspekhi Sovremennogo Estestvoznaniya,* 2004, № 11, p. 65.

[105] Burya A.I., Kozlov G.V., *Rula I.V. Mater. 4th International Conf. "Research and Development in Mechanical Industry"*. RaDMI-2004, 31 Aug.-4 Sept. 2004, Zlatibor, Serbia and Montenegro, p. 334-338.

[106] Berstein V.A., Egorov V.M. *Differential Scanning Calorymetry in Physics-Chemistry of Polymers.* Leningrad, Khimiya, 1990, 256 p.

[107] Burya A.I., Kozlov G.V., Rula I.V. *Novye Materialy i Technologii v Metallurgii i Mashinostroenii,* 2005, № 1, p. 32-36.

[108] Novikov V.U., Kozlov G.V., Yudunov V.V. *Materialovedenie,* 2002, № 4, p. 4-9.

[109] Kozlov G.V., Novikov V.U. *Materialovedenie,* 1997, № 8-9, p. 3-6.

[110] Vilgis T.A. *Physica A,* 1988, v. 153, № 2, p. 341-354.

[111] Ziebland H. In book: Polymer Engineering Composites. Ed. Richardson M.O.W. London, *Applied Science Publishers LTD,* 1978, p. 280-315.

[112] Kozlov G.V., Burya A.I., Zaikov G.E. In book: Molecular and High Molecular Chemistry: Theory and Practice. Ed. Monakov Yu., Zaikov G. New York, *Nova Science Publishers, Inc.,* 2006, p. 131-137.

[113] Burya A.I., Kozlov G.V., Van'kov A.Yu. *Novosti Nauki Pridneprov'ya,* 2004, № 3, p. 33-36.

[114] Yu. Z., Ait-Kadi A., Brisson *J. Polymer Engng. Sci.,* 1991, v. 31, № 16, p. 1222-1227.

[115] Solomko V.P. Filled Crystallizing Polymers. Kiev, *Naukova Dumka,* 1980, 264 p.

[116] Kozlov G.V., Beloshenko V.A., Varyukhin V.N., Lipatov Yu.S. *Polymer,* 1999, v. 40, № 4, p. 1045-1051.

[117] Burya A.I., Kozlov G.V., Chigvintseva O.P. Istraživanje i razvoj, 2005, v. 11, № 3-4, p. 25-31.

[118] Graessley W.W., Edwards S.F. *Polymer,* 1981, v. 22, № 10, p. 1329-1334.

[119] Mandelkern L. *Polymer J.,* 1985, v. 17, № 1, p. 337-350.

[120] Vittoria V. *J. Polymer Sci.: Polymer Phys.* Ed., 1986, v. 24, № 2, p. 451-455.

[121] Seguela R., Rietsch F. *Polymer,* 1986, v. 27, № 5, p. 703-708.

[122] Kozlov G.V., Beloshenko V.A., Varyukhin V.N., Novikov V.U. *Zhurnal Fizicheskikh Issledovanii,* 1997, v. 1, № 2, p. 204-207.

[123] Haward R.N. *Macromolecules,* 1993, v. 26, № 22, p. 5860-5869.

[124] Kozlov G.V., Zaikov G.E. *Izvestiya KBSC RAN,* 2003, № 1(9), p. 54-57.

[125] Krigbaum W.R., Roe R.-G., Smith K.J. *Polymer,* 1964, v. 5, № 3, p. 533-542.

[126] Kozlov G.V., Burya A.I., Dolbin I.V. *Kompozitnye Materialy,* 2008, v. 2, № 1, p. 3-7.

[127] Wunderlich B. Macromolecular Physics. V. 2. *Crystal Nucleation, Growth, Annealing.* New York, San-Francisco, London, Academic Press, 1976, 558 p.

[128] Aloev V.Z., Kozlov G.V., Zaikov G.E. *Russian Polymer News,* 2001, v. 6, № 4, p. 63-65.

[129] Privalko V.P., Kawai T., Lipatov Yu.S. *Polymer J.,* 1979, v. 11, № 9, p. 699-709.

[130] Aloev V.Z., Burya A.I., Kozlov G.V. *Doklady NAN Ukraine,* 2003, № 6, p. 123-127.

Chapter 2

THE ROTATING ELECTROMAGNETIC FIELD INFLUENCE ON CARBON PLASTICS STRUCTURE

At present it is well-known [1, 2], that magnetic interactions, which are negligible by energy, strongly influence on high-energetic processes in condensed mediums. The magnetic principles by physical and chemical processes operation arose, which have no energetic nature, but the spin one [1]. As a result, the new section of different nature processes kinetics occured – spin dynamics, considering spin evolution influence in electron and nuclear subsystems on macroscopic processes in molecular systems and condensed mediums. Studying weak magnetic fields influence on semiconductors an effects whole number was observed, such luminescence arising, mechanical properties change and so on [1]. It has been shown, that dynamics of structural changes in silicium crystals, induced by an impulse magnetic field (IMF), differs in specific features, distinctive for behaviour of spatial-temporal dissipative (synergetic) structures. In papers [1, 2] while studing melting-crystallization kinetics of ionic KCl crystals, beforehand processed by IMF, the appreciable changes of transitional processes nonequilibrium thermodynamic parameters and the major one – melting enthalpy were found, testifying to magnetic field induction of special nonequilibrium phase state in nonmagnetic crystals.

The similar by their physical significance effects were observed for carbon plastics on the basis of phenylone at using components preliminary blending in rotating electromagnetic field technology (see section 1.4). Proceeding from this, the authors [3] studied the interactions polymer-electromagnetic field for the mentioned carbon plastics.

Let's us consider the possible scheme of fractal clusters interaction (let us remind, that polymers amorphous state structure is simulated as a large number set of Witten-Sander clusters, having the average fractal dimension ~ 2.50 [4]) with electromagnetic field. The model [4] supposes that all mentioned clusters are disposed close to each other (the distance between them is about 10 mm) and each of them is capable to feel a neighbouring cluster electromagnetic field [5]. If an interaction field vector for these clusters is directed in the opposite direction in respect to the imposed electromagnetic field vector, then this results to cluster interaction resonance frequency decreasing. Therefore, if ω_0 and ω are initial and obtained owing to interaction frequencies of electron density oscillations of forming a cluster particles, then in the considered case $\omega > \omega_0$. The resonance frequency displacement (ω-ω_0) is proportional to r_{cl}^{-3}, where r_{cl} is the distance between clusters centers.

Let us estimate the distance r_{cl} within the frameworks of the cluster model of a polymers amorphous state structure [6, 7], according to which the cluster radius R_{cl} ($r_{cl}=2R_{cl}$) is given as follows:

$$R_{cl} = 18\left(\frac{2\nu_{cl}}{F}\right)^{1/3}, \text{Å}, \tag{2.1}$$

where ν_{cl} is physical entanglements cluster network density, F is cluster functionality or a number of molecular chains emerging from it.

The value ν_{cl} can be calculated by using the equations (1.1), (1.2), (1.4), (1.16) and (1.88) and functionality F is estimated according to the relationship [7]:

$$F = \frac{4M_{fl}}{M_{cl}}, \tag{2.2}$$

where M_{fl} and M_{cl} are molecular weights of chain section between entanglements traditional nodes (macromolecular "flings" [8]) and clusters, accordingly.

As it is known [8], the value M_{fl} is scaled with C_∞ as follows:

$$\frac{M_{fl}^p}{M_{fl}^c} \sim \left(\frac{C_\infty^p}{C_\infty^c}\right)^2, \tag{2.3}$$

where the indices "*p*" and "*c*" relate to matrix polymer and composite, accordingly.

And at last, M_{cl} value can be calculated according to the following equation [3]:

$$M_{cl} = \frac{\rho N_A S l_0 C_\infty}{\varphi_{cl}}. \tag{2.4}$$

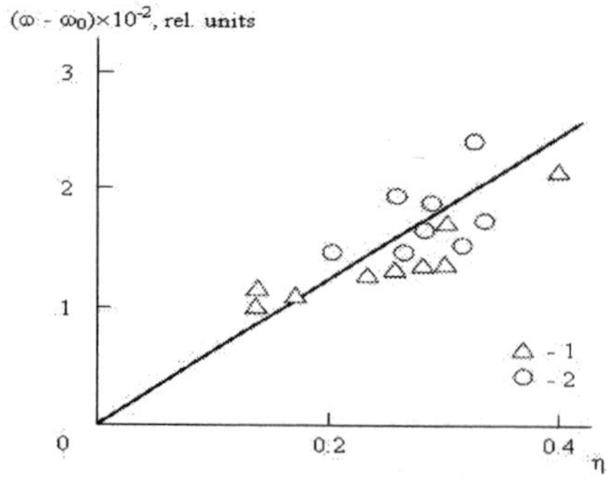

Figure 2.1. The dependence of resonance frequency displacement (ω-ω_0) on fibers orientation factor η for carbon plastics on the basis of phenylone, prepared with magnetic (1) and mechanical (2) separation application [3].

In Figure 2.1 the dependence (ω-ω_0) on carbon plastics synergetic structure governing parameter – fibers orientation factor η [9] is adduced. As one can see, the linear growth (ω-ω_0) at η increasing is observed and at η=0 (ω-ω_0)=0, i.e. frequency of polymer segments electron density oscillations does not change: $\omega=\omega_0$. In other words, at carbon fibers orientation absence field interaction clusters actually are absent [5].

As it is known [5], the electromagnetic wave absorption coefficient K_{em} for consisting of fractal clusters substance is given as follows:

$$K_{em} = K_{em}^0 m^{(3/d_f)-1}, \tag{2.5}$$

where K_{em}^0 is an absorption coefficient for Euclidean substance at d_f=d=3, m is cluster mass, d_f is its fractal dimension.

As m statistical segments number n_{cl} in one cluster should be accepted. Since cluster is an amorphous analog of crystallite with the extended chains, then it can be written [7]:

$$n_{cl} = \frac{F}{2}. \tag{2.6}$$

In Figure 2.2 the dependence K_{em} on duration of components blending in rotation electromagnetic field t for the studied carbon plastics is shown, which allows to make two conclusions. At first, this dependence has synergetic character: in the range t=5-120 s the approximately sigmoid dependence $K_{em}(t)$ is obtained, which at $t\geq120$ s approaches to the constant K_{em} value. As it is known [10] (see section 1.1 also), such dependence type is specific for periodic (quasiperiodic) systems with subsequent structure transition to chaotic behaviour. Secondly, the values K_{em} for carbon plastics, prepared with mechanical separation application, are approximately smaller on 20 % than the corresponding magnitudes for samples, prepared with magnetic separation application. This result supposes that the remaining in composition ferromagnetic particles wear products in case of mechanical separation using screen or partly compensate rotating electromagnetic field action.

As it is shown in paper [11], electromagnetic field action on macromolecular coil can be described by the equation:

$$f = \frac{3\langle R_g\rangle}{l_{st}N}, \tag{2.7}$$

where f is acting to coil force, which is due to electromagnetic field availability, $\langle R_g\rangle$ is meansquare gyration radius of macromolecular coil, l_{st} is statistical segment length, N is polymerization degree.

Besides, for the same parameters it can be written [12]:

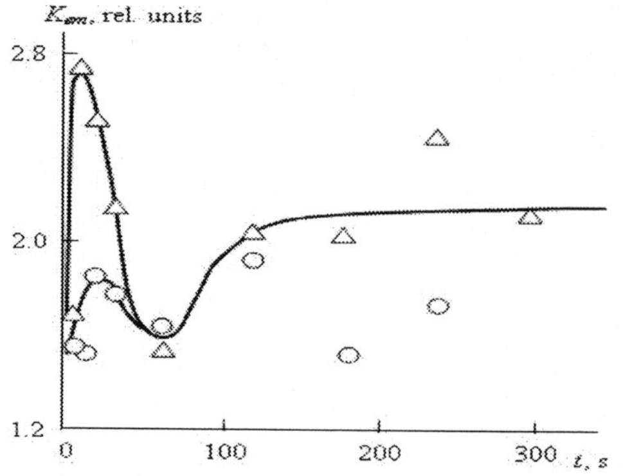

Figure 2.2. The dependence of absorption coefficient K_{em} on duration of components blending in rotating electromagnetic field t for carbon plastics on the basis of phenylone. The designations are the same as in Figure 2.1 [3].

$$C_\infty = \frac{\langle R_g^2 \rangle}{n l_0}, \tag{2.8}$$

where n is a main chain real or virtual bonds number.

Then, assuming $f \sim K_{em}$ and accounting for the equation (1.14), let us obtain [3]:

$$C_\infty \sim K_{em}^{-2}. \tag{2.9}$$

Plotting of the dependence, corresponding to the relationship (2.9), confirms this correlation. It is significant that at K_{em}=0 the value C_∞ has the smallest magnitude 2.0, i.e. it corresponds to the chain with tetrahedral valent angles [12]. From these data it follows, that absorption coefficient K_{em} increasing results to macromolecular coil compactness decreasing, i.e. to R_g growth (the equation (2.8)).

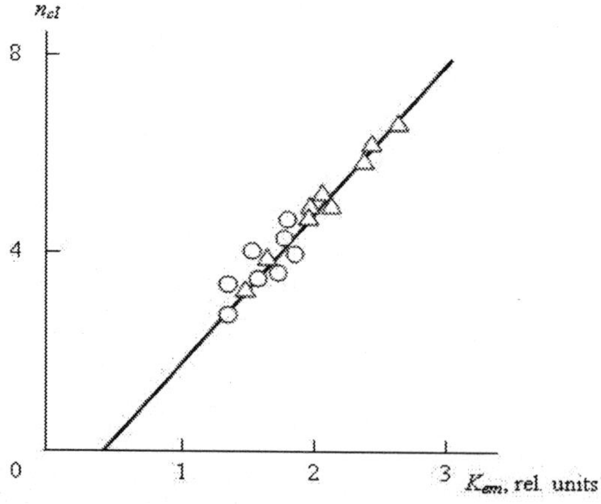

Figure 2.3. The dependence of statistical segments number in one cluster n_{cl} on absorption coefficient K_{em} for carbon plastics on the basis of phenylone. The designations are the same as in Figure 2.1 [3].

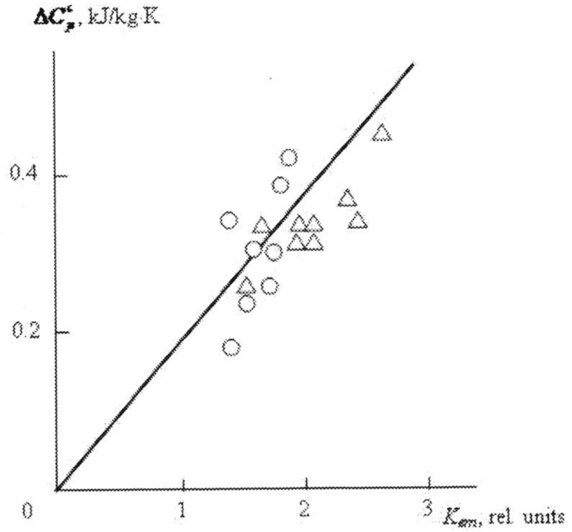

Figure 2.4. The dependence of heat capacity jump at constant pressure at glass transition temperature ΔC_p^c on absorption coefficient K_{em} for carbon plastics on the basis of phenylone. The designations are the same as in Figure 2.1 [3].

In Figure 2.3 the dependence $n_{cl}(K_{em})$ is shown, from which the growth of statistical segments in one cluster at intensification of electromagnetic wave absorption by polymer follows. It is significant that at $K_{em}\approx0.8$ relative units the value $n_{cl}=1.0$, i.e. in this case clusters in polymeric matrix are not formed. The dependence of heat capacity jump at constant pressure at glass transition temperature ΔC_p^c of carbon plastics on K_{em}, adduced in Figure 2.4, allows to make the similar conclusion. In this case K_{em} growth results to ΔC_p^c increasing. In its turn, the interfacial regions relative fraction φ_{if} can be calculated according to the equation (1.11), from which it follows, that at $K_{em}=0$ $\Delta C_p^c=0$ and $\varphi_{if}=1.0$, i.e. polymeric matrix structure will be identical to interfacial layer structure. At $\Delta C_p^c=\Delta C_p^p$ the value $\varphi_{if}=0$, that can be realized at $K_{em}\approx2.8$ relative units.

The clusters formation in polymeric matrix defines its fractal dimension d_f (see the equation (1.4)) and the dependence $d_f(K_{em})$ adduced in Figure 2.5, confirms this. The electromagnetic wave absorption intensification by clusters results to n_{cl} increasing (Figure 2.3) and φ_{cl} raising, that decreases the value d_f. The mentioned above limits $K_{em}=0.8$-2.8 relative units give the range of values $d_f=2.28$-2.73, that is completely confirmed by both theoretical estimations [13] and experimental data (Figure 1.1).

And at last, let us principally note a very important fact. Clusters and electromagnetic field interaction indicates quantum properties availability for them and, hence, on their belonging to nanoworld [14]. Clusters self assembly mechanism from statistical segments and clusters size allows to make such conclusions. Let us make the following estimations, using the adduced above data. One can get the greatest value $n_{cl}=m=28$ from the equation (2.5) at the greatest value $K_{em}=2.8$ relative units and at the smallest magnitude $d_f=2.28$. Further having calculated statistical segment volume as Sl_{st} and V_{cl} as $Sl_{st}n_{cl}$ one can get cluster volume $V_{cl} \approx 2.46 \times 10^{-27}$ m^3. The cluster molecular weight M_{cl} can be calculated according to the equation (1.36) that gives $M_{cl} \approx 2100$. Assuming atom average mass as equal to ~ 15, let us obtain atoms number in cluster ~ 140. This value is lower than the upper limit of 10^3-10^4 atoms for nanoparticles, calculated in paper [15]. The indicated observations allow to consider clusters as nanoparticles.

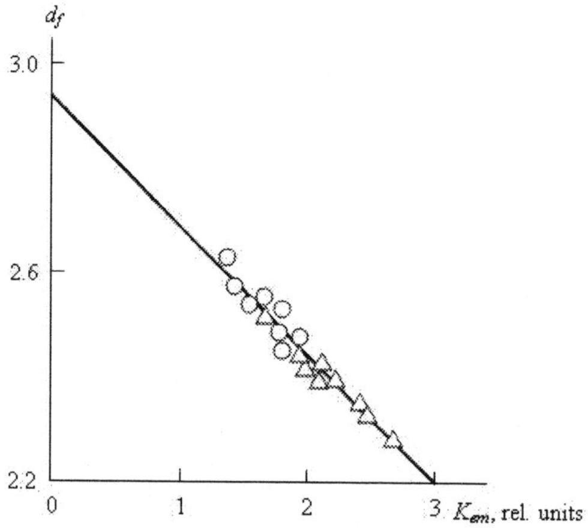

Figure 2.5. The dependence of polymeric matrix structure fractal dimension d_f on absorption coefficient K_{em} for carbon plastics on the basis of phenylone. The designations are the same as in Figure 2.1 [3].

Let us note in conclusion, that electromagnetic field influence on polymers according to the signs number (field influence on nonmagnetic materials, electromagnetic field postaction) is completely similar to IMF influence on crystalline low-molecular materials.

Therefore, the stated above results demonstrated essential influence of rotating electromagnetic field on carbon plastics polymeric matrix structure formation. The intensification of electromagnetic wave absorption by polymer results to local order domains (clusters) size increase, bulk polymeric matrix fractal dimension reduction and interfacial regions relative fraction decreasing. According to the signs number one should attribute clusters to nanoparticles.

In papers [1, 2] it was also shown, that preliminary processing of ionic and covalent crystals by IMF substantially influences on their melting process at the expence of thermodynamic parameters essential change including melting enthalpy. This can be considered as the evidence of special nonequilibrium structure in nonmagnetic crystals induction by electromagnetic field [2].

Practically such changes were found in carbon plastics on the basis of phenylone, prepared by the technology of components preliminary blending in rotating electromagnetic field. Depending on blending duration t (t=5-300 s) carbon plastics with essentially differing properties at the same carbon fibers contents can be obtained. So, for example, their elasticity modulus can be changed from 2.13 up to 3.33 GPa, strength at compression – from 294 up to 406 MPa and at impact toughness – from 3.2 up to 19.2 kJ/m^2.

Therefore in paper [16] the study of vitrification process kinetics and structure of carbon plastics is fulfilled on the basis of phenylone, prepared by technology of preliminary components blending in rotating electromagnetic field.

In Figure 2.6 the dependence of vitrification enthalpy change ΔH on duration of components blending in rotating electromagnetic field t for the studied carbon plastics is shown. As one can see, at t change within the range of 5-300 s the essential changes of ΔH value from 3.42 up to 8.76 kJ/kg occur. The dependence $\Delta H(t)$ has again a typical synergetic character: within the range of t=5-120 s vitrification enthalpy changes almost according to the sigmoid law and at $t \geq 120$ s the value ΔH becomes constant and equal to ~ 7 kJ/kg. As it is known [10], such dependences type is specific for periodic (quasiperiodic) structures with subsequent system transition to chaotic behaviour, that supposes substantial change of supermolecular (more precisely, supersegmental) thermodynamically nonequalibrium structure of polymeric matrix under rotating electromagnetic field influence. One can imply that in the considered carbon plastics, as well as in nonmagnetic crystals [1, 2], the postaction ("memory") effect is possible owing to which samples preparation by hot pressing method does not result to levelling (erasing) of structural changes, which occur at components blending in rotating electromagnetic field.

At present it is accepted to consider, that glass transition (vitrification) is not the phase transition and consequently thermal-physical parameters change at glass transition temperature T_g characterizes a nonequilibrium transition, at which heat capacity is consumed not on structure major changes, but on polymer chains repacking at transition to less nonequilibrium states [17]. The glass transition heat (enthalpy) ΔH in this case is defined as follows:

$$\Delta H = \int_{T_1}^{T_2} \Delta C_p(T) dT , \tag{2.10}$$

where T_1 and T_2 are glass transition beginning and end temperatures (Figure 2.7) and ΔC_p is a heat capacity jump at constant pressure at temperature T_g.

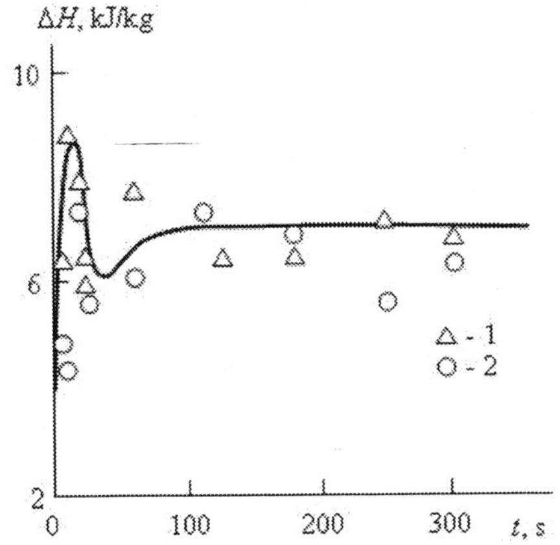

Figure 2.6. The dependence of vitrification enthalpy ΔH on duration of components blending in rotating electromagnetic field t for carbon plastics on the basis of phenylone, prepared with magnetic (1) and mechanical (2) separation application [16].

As it was noted above, the substance structural state universal characteristic is fractal dimension d_f. Consequently in Figure 2.8 the dependence $d_f(\Delta H)$ is adduced, from which d_f increasing at ΔH reduction follows where fractal dimension changes within the limits of 2.28-2.67 [16]. As it was shown above, d_f variation supposes that rotating electromagnetic field influences on all structural levels of amorphous phenylone – molecular, topological and supersegmental [16].

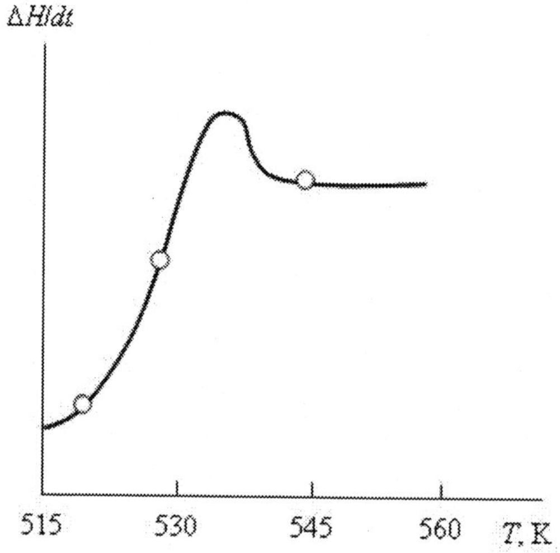

Figure 2.7. The schematic representation of the temperature dependence of enthalpy change rate $\Delta H/dt$ for phenylone [16].

There is an obvious analogy in electromagnetic field influence on ionic and covalent (nonmagnetic) crystals, on the one hand, and on amorphous phenylone, on the other hand:

- in both cases electromagnetic field influences on nonmagnetic materials, that supposes this influence spin nature [1];
- the postaction effect is fulfilled, i.e. electromagnetic field action is not removed by the subsequent processing ("memory" effect) [2];
- an electromagnetic field influences essentially on nonequilibrium thermodynamic transitions parameters;
- the electromagnetic field action results to essentially different nonequilibrium structure formation, that for amorphous polymers is expressed in the changes on all structural levels at invariable chemical constitution of polymers;
- changes dynamics has a synergetic character, typical for spatial-temporal dissipative structures behaviour.

Consequently, the experimental data by glass transition enthalpy change of carbon plastic on the basis of phenylone showed that rotating electromagnetic field imposition at composite components blending influenced strongly on matrix amorphous phenylone structure, which was displayed on molecular, topological and supersegmental structural levels [16].

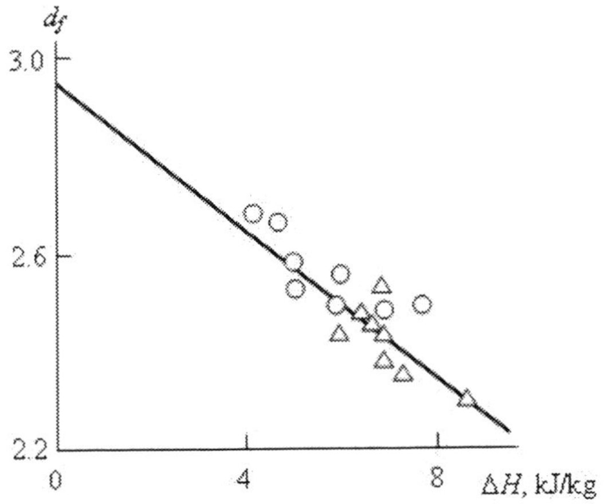

Figure 2.8. The dependence of structure fractal dimension d_f on glass transition enthalpy ΔH for carbon plastics on the basis of phenylone. The designations are the same as in Figure 2.6 [16].

According to Kadomtsev-Shevchenko synergetic conception [18] nanoworld separates macroworld from elementary particles world, nanoworld objects have classical, quantum and principally new properties at that. Earlier on the basis of analysis of structure formation near low-molecular substances melting temperature T_m effects the temperature range $T^{'} \leq T_m \leq T^{''}$ was distinguished, which characterizes structures self-organization not characteristic to the first kind of phase transition [1, 2]. In connection with nanoworld special properties conception this transitional range availability should be connected with the range, characterized by nanocluster structures self-organization; it separates structures with specific features for solid and liquid states. The similar dependence also is observed at amorphous polymers glass transition (Figure 2.7), which transition is due to local order domains (clusters) formation, having sizes of nanometer range (~ 0.5-2.0 nm) [6, 7].

For studing the dependence of glass transition temperature T_g on particles (statistical segments) number in one cluster n_{cl} the authors [19, 20] used carbon plastics on the basis of phenylone, prepared by the technology of components preliminary blending in rotating electromagnetic field. The values T_g were determined by a differential scanning calorimetry method and the values n_{cl} were calculated according to the equations (2.2) and (2.6). In Figure 2.9 the dependence $T_g(n_{cl})$ is shown, from which T_g reduction at n_{cl} growth follows. The similar dependence $T_m(n_{cl})$ was obtained for silicium nanoclusters [2].

By analogy with the equation (1.17) the numbers ratio

$$A_m = \frac{n_{cl}^{i}}{n_{cl}^{i+1}} = \Delta_i^{1/m} \qquad (2.11)$$

characterizes nanocluster structure adaptivity A_m to temperature increase, determined by its stability measure Δ_i and feedback parameter m (at m=1 the linear feedback is realized and at m≥2 – the nonlinear one) [10]. From the data, adduced in table 2.1, it follows that the line “b” in Figure 2.9 is related to clusters with lower stability (Δ_i=0.213) in comparison with the line "a", corresponding to clusters with higher glass transition temperature (Δ_i=0.324). Therefore, the temperature dependence of a statistical segments number in stable cluster depends on packing mechanism (the line "a" and m=2; 4 correspond to dense packing; the line "b" and m=64 – to the loose one) [20].

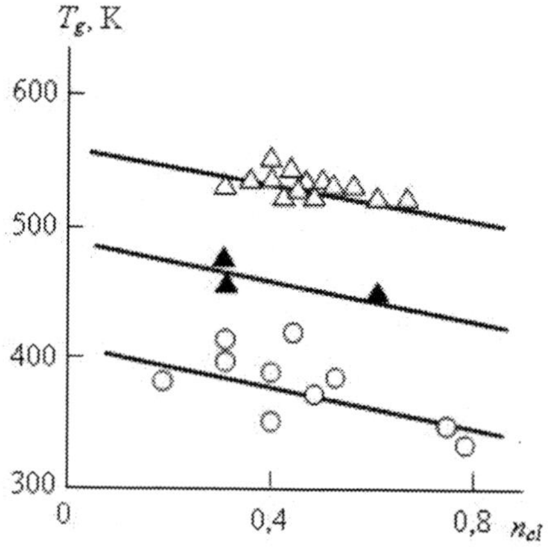

Figure 2.9. The dependence of glass transition temperature T_g on statistical segments number in one cluster n_{cl} for carbon plastics on the basis of phenylone. The explanations are given in the text [20].

As it is known [7], the value T_g of both amorphous polymers and composites on their basis is defined by a structure order parameter φ_{cl} and this parameter increase results to T_g growth. However, the adduced in Figure 2.10 correlation $T_g(\varphi_{cl})$ for the studied carbon plastics shows such dependence absence and for Δ_i=0.213 T_g≈const≈460 K and for Δ_i=0.324 T_g≈const=530 K. Therefore, rotating electromagnetic field application creates mesophase (cluster structure) with principally new properties. Let us note, that T_g increase at Δ_i growth assumes their following interconnection [20]:

$$T_g = 293 + 790\Delta_i , \qquad (2.12)$$

Table 2.1. The synergetic characteristics of nanoclusters formation for phenylone [20]

The point number in Figure 2.9	1	2	3	4	5
n_{cl}	2.76	4.94	6.62	3.40	3.51
A_m	-	0.559	0.746	-	0.959
Δ_i	-	0.324	0.324	-	0.213
m	-	2	4	-	64

where T_g and constants in the equation (2.12) right-hand part are given in absolute degrees.

At Δ_i=0 (n_{cl}^{i}=0, i.e. in case of clusters collapse) T_g is equal to the testing temperature (in the considered case – to 293 K), as and should be expected for rubbers. Since the greatest value Δ_i=0.618 at m=1 [10], then the equation (2.12) allows to estimate the greatest possible value T_g which is equal to 781 K for the studied composites. The equation (2.12) within the frameworks of synergetics confirms the cluster model postulate [7]: the value T_g is defined by not only (and not so much) polymer chemical nature, but also by its structure.

Therefore, the adduced above analysis of structure formation at glass transition with due regard to nanoworld synergetic conception allows to make the conclusion, that glass transition temperature is a bifurcation point, corresponding to local order domains (clusters) degeneration and rotating electromagnetic field application creates mesophase (cluster structure) with principally new properties [19, 20].

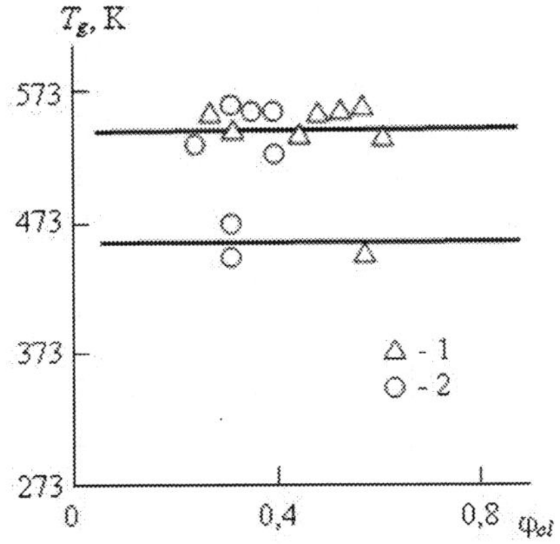

Figure 2.10. The dependence of glass transition temperature T_g on clusters relative fraction φ_{cl} for carbon plastics on the basis of phenylone, prepared with magnetic (1) and mechanical (2) separation application [20].

REFERENCES

[1] Mashkina E.S. Mater. of International Interdisciplinary Symposium "Fractals and Applied Synergetics, FaAS-03". Moscow, *Publishers MSOU,* 2003, p. 297-298.

[2] Bityutskaya L.A., Mashkina E.S., Butusov I.Yu. *Pis'ma v ZhTF,* 2001, v. 27, № 20, p. 14-19.

[3] Kozlov G.V., Burya A.I., Dolbin I.V. *Prikladnaya Fizika,* 2006, № 1, p. 14-18.

[4] Shogenov V.N., Kozlov G.V. Fractal Clusters in Physics-Chemistry of Polymers. Nal'chik, *Polygraphservice and T,* 2002, 270 p.

[5] Zolotukhin I.V., Kalinin Yu.E., Stogney O.V. A New Directions of Physical Material Science. Voronezh, *Publishers VSU,* 2000, 360 p.

[6] Kozlov G.V., Novikov V.U. *Uspekhi Fizicheskikh Nauk,* 2001, v. 171, № 7, p. 717-764.

[7] Kozlov G.V., Zaikov G.E. Structure of the Polymer Amorphous State. Utrecht-Boston, *Brill Academic Publishers,* 2004, 465 p.

[8] Wu S. J. Polymer Sci.: Part B: *Polymer Phys., 1*989, v. 27, № 4, p. 723-741.

[9] Burya A.I., Kozlov G.V., Kholodilov O.V. *Vestnik Polotskogo Gosuniversiteta, seriya B,* 2005, № 6, p. 36-39.

[10] Ivanova V.S., Kuzeev I.R., Zakirnichnaya M.M. Synergetics and Fractals. Universality of Materials Mechanical Behaviour. Ufa, *Publishers USNTU,* 1998, 366 p.
[11] Kholodenko A., Vilgis T.A. Phys. Rev. E, 1994, v. 50, № 2, p. 1257-1264.
[12] Budtov V.P. *Physical Chemistry of Polymers Solutions.* Sankt-Peterburg, Khimiya, 1992, 384 p.
[13] Kozlov G.V., Burya A.I. *Doklady NAN Ukraine,* 2006, № 3, p. 136-142.
[14] Folmanis G.E. Mater. of International Interdisciplinary Symposium "Fractals and Applied Synergetics, FaAS-03". Moscow, *Publishers MSOU,* 2003, p. 303-308.
[15] Shevchenko V.Ya., Bal'makov M.D. *Fizika i Khimiya Stekla,* 2002, v. 28, № 6, p. 631-636.
[16] Kozlov G.V., Burya A.I., Shustov G.B. *Fizika i Khimiya Obrabotki Materialov,* 2005, № 5, p. 81-84.
[17] Bartenev G.M., Frenkel' S.Ya. *Physics of Polymers.* Leningrad, Khimiya, 1990, 432 p.
[18] Kadomtsev B.B. *Dynamics and Information. Moscow,* Publishers zhurn. UFN, 1999, 272 p.
[19] Dolbin I.V., Burya A.I., Kozlov G.V. *Proceedings of XX International Conf. "Energy Intensive Streams Influence on Matter".* El'brus, KBSU, 2005, p. 35.
[20] Dolbin I.V., Burya A.I., Kozlov G.V. In book: The Physics of Substance Extreme States". Ed. Fortov V. Chernogolovka, *IFKh,* 2005, p. 67-69.

Chapter 3

MECHANICAL PROPERTIES OF POLYMER COMPOSITES FILLED WITH SHORT FIBERS

As it is known [1], the main aim of polymers physics is quantitative relationships structure-properties obtaining. For polymer composites, filled with short fibers, two research directions are of undoubted interest. At first, it is the research of carbon plastics on the basis of phenylone, which have, as it was noted above, mechanical properties essential variation at filler constant content. This supposes corresponding change of carbon plastics structure, owing to which they are a suitable object for relationships structure-properties obtaining. The second studies direction can be considered as the classical one: this is a mechanical properties change investigation at filler contents variation. Both indicated directions will be presented in the present chapter.

3.1. THE ELASTICITY MODULUS AND COMPOSITES REINFORCEMENT MECHANISMS

The elasticity modulus E_c is one the most important mechanical characteristics of polymer composites. Often filler introduction in polymer is defined by considerations of high-modulus polymer material preparation [2]. Within the frameworks of continuous approach a large number of models exists, giving the value E_c quantitative description as a function of polymeric matrix and filler characteristics [3]. Since all these models were elaborated with reference to some concrete case, they were not able to the give an answer to general questions of composites structure: to describe the structure of interfacial boundaries polymer-filler, distribution, size and shape of filler particles and so on [3]. Therefore lately for the indicated purposes modern physical conceptions (synergetics, fractal analysis, irreversible aggregation models) are often used more [4, 5]. So, within the frameworks of fractal analysis for particulate-filled polymer composites it has been shown, that the value E_c increases linearly at the excess energy localization regions dimension D_f raising, which "pumped" into polymeric matrix, that is expressed analytically as follows [5]:

$$E_c = c(D_f - 3), \text{ GPa}, \tag{3.1}$$

where c is the constant, which is equal to ~ 0.67.

Within the frameworks of such approach it has been postulated that the value E_c is defined only by polymeric matrix structure and the filler role comes to this structure modification ("disturbance") and its subsequent fixation. The authors [6] fulfilled this postulate verification and obtained the elasticity modulus quantitative description for carbon plastics on the basis of phenylone.

In Figure 3.1 the dependence of elasticity modulus E_c on square root of D_f, determined according to the equation (1.21) (such dependence form was selected for its linearization) is adduced. As one can see, in this case the good linear correlation is actually obtained, expressed analytically as follows [6]:

$$E_c = 3.70\left(D_f^{1/2} - 1\right), \text{ GPa.} \qquad (3.2)$$

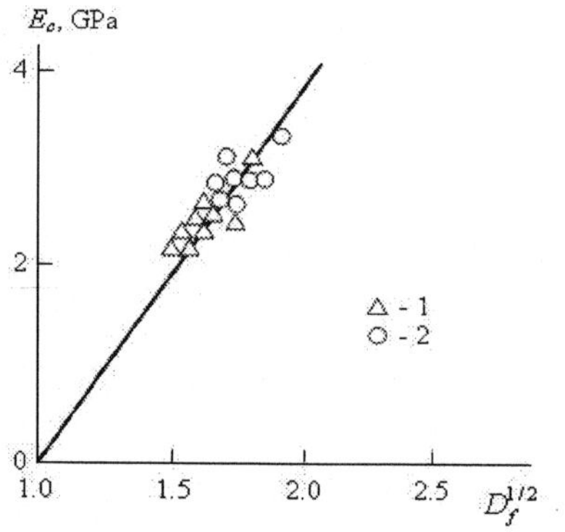

Figure 3.1. The dependence of elasticity modulus E_c on the excess energy localization regions dimension D_f for carbon plastics on the basis of phenylone, prepared with magnetic (1) and mechanical (2) separation application [6].

One of the fractal analysis values is a clear definition of limiting magnitudes of its main characteristics – fractal dimensions. This gives possibility to estimate the limiting values E_c for the studied carbon plastics. Poisson's ratio ν the greatest value for real solids is equal to 0.475 [7], the greatest value D_f=21 is obtained from the equation (1.21)and from the equation (3.2) – the limiting greatest magnitude E_c=13.3 GPa. Let us note, that although for carbon plastics the dependence E_c on D_f is expessed weaker, than for the particulate-filled composites (compare the equations (3.1) and (3.2)), but the constant coefficient distinction in these equations results to the higher maximum value E_c in comparison with particulate-filled composites (E_c=13.3 and 12.1 GPa, accordingly). For ν=0 D_f=2.0 and the smallest value E_c=1.53 GPa for carbon plastics, that is approximately equal to matrix polymer elasticity modulus. And at last, for strongly porous composites ν=-1.0, D_f=1.33 and E_c=1.23 GPa. Let us note, that for particulate-filled composites the lower limiting value D_f=3 or ν=0.25, that is the condition of material complete brittleness reaching [7], when E_c=0 according to the equation (3.1). It is significant that for carbon plastics the condition D_f=1 and, consequently, E_c=0 is unattainable, that follows from the equation (1.21). The distinction of limiting characteristics, i.e. D_f and E_c, for polymer composites, filled with particulates and short fibers, requires special study.

As it is known [8], carbon plastics structure includes two densely-packed components: local order domains (clusters) in bulk polymeric matrix and interfacial regions with relative fractions φ_{cl} and φ_{if}, accordingly. Besides, this structure is a synergetic system, possessing feedback [9], in virtue of which between parameters φ_{cl} and φ_{if} interconnection exists, expressed by the relationship (1.10). Since this relationship shows that the physical significance of feedback for carbon plastics structure is contained in polymeric material "pumping" from one densely-packed structural component into another one then it one can suppose that the possible greatest structure equilibrium is realized at one of the components prevailence or the greatest difference $|\varphi_{cl}-\varphi_{if}|$. Actually, the adduced in Figure 3.2 dependence of D_f on difference $|\varphi_{cl}-\varphi_{if}|$ absolute value has shown D_f reduction at the indicated difference growth, i.e. energy, "pumped" in polymeric matrix, decrease or polymeric matrix structure "disturbance" degree reduction. According to the equation (3.2) this means E_c decrease at $|\varphi_{cl}-\varphi_{if}|$ growth. From the data of Figure 3.2 it follows, that at the densely-packed structural components the greatest fraction, which is equal to 0.74 [8], the value D_f=2.25 or from the equation (3.2) the lower limiting value E_c for real carbon plastics can be obtained, which is equal to 1.85 GPa [6].

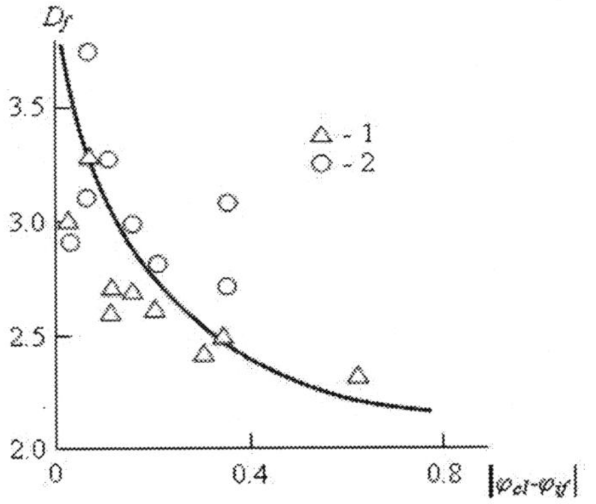

Figure 3.2. The dependence of excess energy localization regions dimension D_f on absolute value of densely-packed structural components relative fractions difference $|\varphi_{cl}-\varphi_{if}|$ for carbon plastics on the basis of phenylone. The designations are the same as in Figure 3.1 [6].

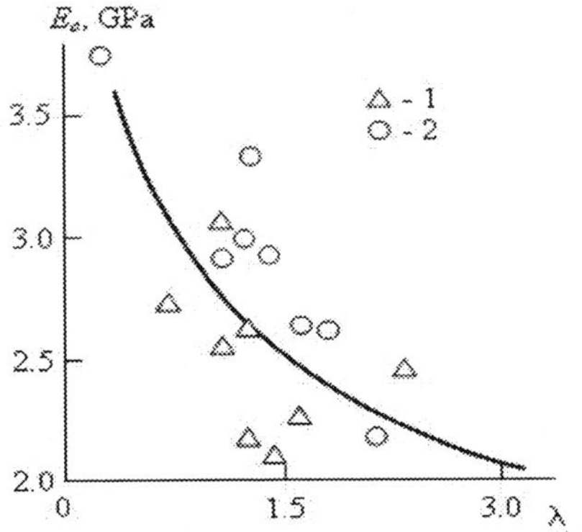

Figure 3.3. The dependence of elasticity modulus E_c on feedback parameter λ for carbon plastics on the basis of phenylone. The designations are the same as in Figure 3.1 [6].

The dependence, adduced in Figure 3.2, supposes certain interconnection of E_c and structure feedback parameter λ, which can be determined with the aid of Puancare equation (1.12), in which the fibers orientation factor η is a governing parameter [10]. In Figure 3.3 the dependence $E_c(\lambda)$ is adduced, which has the expected character, although with large enough scatter. λ increase means polymeric material "pumping" intensification in one of the densely-packed structural components, D_f decrease (Figure 3.2) and, as consequence, E_c reduction (Figure 3.1 and the equation (3.2)). Let us note, that the entire considered variation E_c is obtained at filler constant contents that emphasizes structural changes role in this parameter definition.

Hence, the stated above results showed the possibility of essential variation of elasticity modulus at filler constant contents at the expence of structural changes. The "pumping" into polymeric matrix energy or its "disturbance" degree increase, characterized by the excess energy localization regions dimension raising, results to elasticity modulus growth and feedback in carbon plastics structure intensification, meaning "pumping" of polymeric material from one densely-packed structural component into another, defines its value reduction [6].

Therefore, it was shown above that although for carbon plastics on the basis of phenylone the dependence $E_c(D_f)$ exists, but in case of compression testing it is expressed by the quadratic relationship (3.2), but not linear one, as for particulate-filled composites in bending testing (see the equation (3.1)). Therefore the authors [11] studied the reasons of the dependences $E_c(D_f)$ different character in the indicated types of tests for polymer composites filled with particulates and short fibers.

The yield stress value at tension σ_Y^t can be calculated by the analogous values at compression σ_Y^{comp} according to the equation [12]:

$$\frac{\sigma_Y^t}{\sigma_Y^{comp}} = \frac{\left(1+\nu^2\right)^{1/2} - \nu}{\left(1+\nu^2\right)^{1/2} + \nu}. \tag{3.3}$$

The equation (3.3) accounts for the fact that Poisson's ratio ν is structural characteristic independent on a loading scheme [7]. Further the elasticity modulus value at tension can be calculated according to the equation (1.2) at the condition $\sigma_Y = \sigma_Y^t$. Besides, the relation between E and σ_Y within the frameworks of anharmonicity conception is described as follows [13]:

$$\sigma_Y = \frac{E}{6\gamma}, \tag{3.4}$$

where γ is Grüneisen parameter of intermolecular bonds, which is connected with Poisson's ratio by the equation [14]:

$$\gamma = A\frac{(1+\nu)}{(1-2\nu)}. \tag{3.5}$$

In the equation (3.5) the coefficient A varies within the limits of 0.7-0.9 [14]. Assuming A=0.9, the authors [5] obtained the following relationship from the combination of the equations (3.4), (1.2) and (3.5):

$$E_c = (8.1D_f - 10.8)\sigma_Y. \tag{3.6}$$

In Figure 3.4 the comparison of the experimental and calculated according to the equation (3.6) dependences $E_c(D_f)$ at tension and compression for both series of the studied carbon plastics is adduced. As one can see, if in the tension case good correspondence of theory and experiment was obtained, then in the compression case these dependences differ not only quantitatively, but also qualitatively: the equation (3.6) gives E_c~D_f and the experimental curve – E_c~ $D_f^{1/2}$.

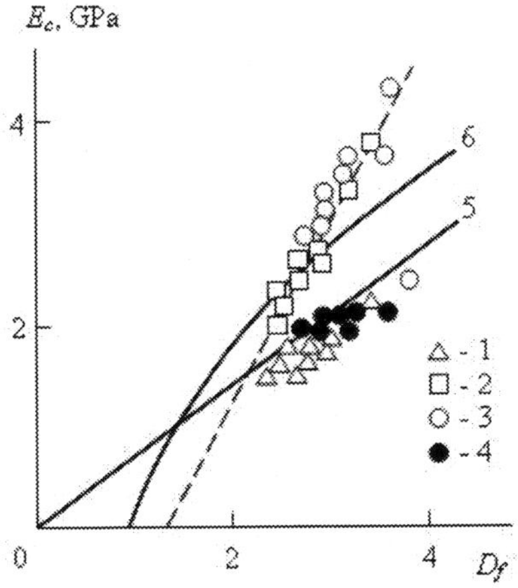

Figure 3.4. The dependences of elasticity modulus E_c on the excess energy localization regions dimension D_f in tension (1, 2, 5) and compression (3, 4, 6) tests for carbon plastics on the basis of phenylone, prepared with magnetic (1) and mechanical (2) separation application. 1-4 – calculation to the equation (3.6); 5, 6 – experimental data [11].

One of the possible reasons of theory and experiment discrepancy can be the usage of the approximation A=const=0.9 at the equation (3.6) derivation [5]. Strictly speaking [14], the coefficient A value is a function of relative fluctuation free volume f_g, that is expressed by the following equation:

$$A = \frac{2}{9}\ln\left(\frac{1}{f_g}\right). \tag{3.7}$$

The value f_g can be estimated according to the equation (1.8). Further, if to suppose, that constant coefficient 8.1 in the equation (3.6) is variable, then it can be designated as K_1 and calculated from this equation, using the experimental magnitudes E_c and σ_Y, obtained in compression tests, and also D_f values, determined according to the equation (1.21). In Figure 3.5 the correlation between K_1 and A is shown, which turns out to be linear and is analytically expressed by the following empirical equation [11]:

$$K_1 = 10.8A. \tag{3.8}$$

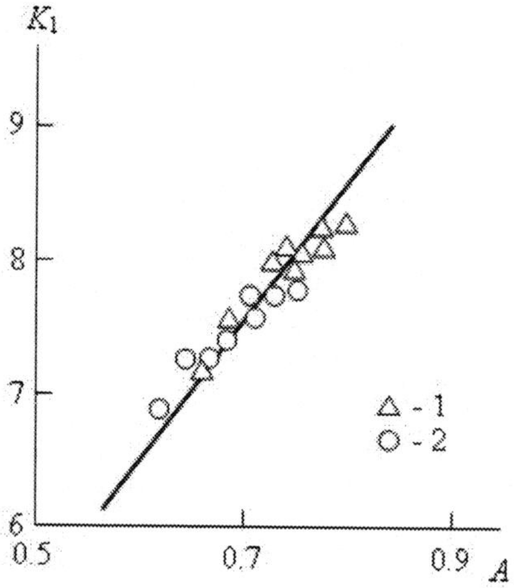

Figure 3.5. The relation between coefficients K_1 and A (the explanations are given in the text) for carbon plastics on the basis of phenylone, prepared with magnetic (1) and mechanical (2) separation application in compression tests [11].

The equation (3.8) substitution in the relationship (3.6) allows to obtain more precise variant of the latter [11]:

$$E_c = 10.8\sigma_Y\left(AD_f - 1\right). \tag{3.9}$$

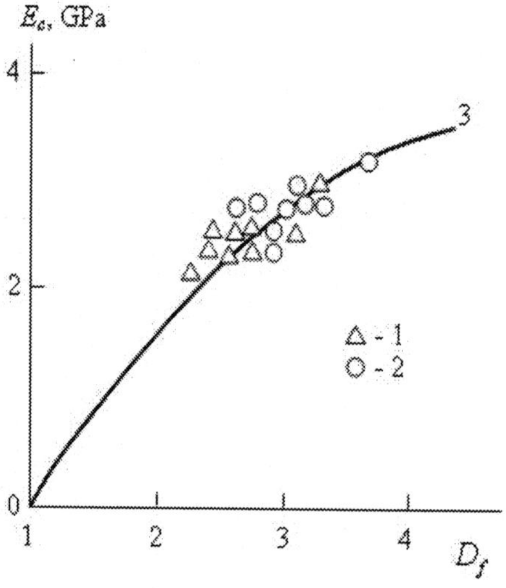

Figure 3.6. The comparison of the experimental (1, 2) and calculated according to the equation (3.9) (3) dependences of elasticity modulus E_c on the excess energy localization regions dimension D_f in compression tests for carbon plastics on the basis of phenylone, prepared with magnetic (1) and mechanical (2) separation application [11].

In Figure 3.6 the comparison of the experimental and calculated according to the equation (3.9) dependences $E_c(D_f)$ for compression tests is adduced. As one can see, now an excellent correspondence of theory and experiment is obtained (the discrepancy is less then 3 %). The equations (3.6) and (3.9) comparison indicates to the essentially differing role of the fluctuation free volume in tension and compression tests. If in the first case f_g role is insignificant and therefore the approximation A=const gives a good correspondence of theory and experiment (Figure 3.4), then in compression case such approximation usage at f_g variation within the limits of 0.027-0.060 can result to the essential (up to 40 %) experimental and theoretical values E_c discrepancy. The fact that f_g increase according to the equation (3.7) reduces E_c at the other equal conditions supposes that in compression tests in the first place just fluctuation free volume microvoids deformation occurs and only after this carbon plastics structure is elastically deformed.

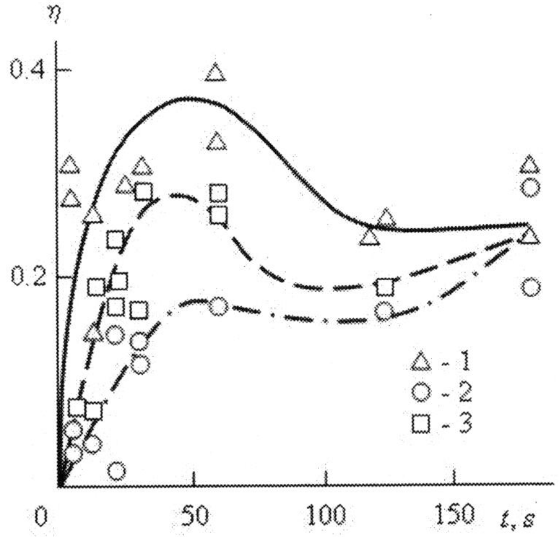

Figure 3.7. The dependence of fibers orientation factor η on components blending duration in rotating electromagnetic field t for carbon plastics on the basis of phenylone with ferromagnetic particles of length 20 (1), 40 (2) and 70 (3) mm application [15].

Let us note in conclusion, that the equation (3.1) was obtained for a bending loading scheme. Such scheme application for carbon plastics with regard that σ_Y and E_c values at bending are equal to half-sum of analogous values at tension and compression, gives the linear relationship between E_c and D_f of form [11]:

$$E_c = 1.40(D_f - 1),\ \text{GPa}. \tag{3.10}$$

Besides, for the bending case a good enough correspondence (the average discrepancy ~ 10 %) of the experimental and calculated according to the equation (3.6) values E_c was obtained.

Hence, the adduced above data showed essentially differing role of fluctuation free volume at elastic deformation in the tension and compression cases. If in the first case its role is insignificant and the acceptance of approximation f_g=const gives theory and experiment permissible discrepancy, then in the compression case the usage of exact values f_g or coefficient A in the equation (3.7) is required and the indicated above approximation can give large discrepancy (up to 40 %) of theory and experiment [11].

It has been shown earlier that carbon fibers in polymeric matrix distribution change, defining carbon plastics structure synergetic character, is realized at the expence of nonequiaxial ferromagnetic particles rotation in electromagnetic field. The carbon plastics, prepared by using ferromagnetic particles of length l_f=40 mm were studied up to now. However, besides them particles with length l_f=20 and 70 mm were applied. Therefore the question about the value l_f influence on carbon plastics structure change as a function of components blending duration in rotating electromagnetic field t arises that was purpose of the present paper [15]. This question is of interest from the technological point of view too as the question of optimal value l_f choice in carbon plastics preparation process.

In Figure 3.7 the dependence of carbon plastics structure governing parameter (fibers orientation factor) η on components blending duration in rotating electromagnetic field t for three using lengths of nonequiaxial ferromagnetic particles is adduced. As one can see, the curves $\eta(t)$ shape for all three l_f is the same: in periodic (ordered) behaviour region the extreme η chaotic behaviour – reaching asymptotic value $\eta \approx 0.25$, which is the same for all three using l_f values. In the range t<180 s absolute values η are l_f a clearly expressed function: the smallest η are obtained for l_f=20 mm, the greatest ones – for l_f=40 mm, whereas η for l_f=70 mm have intermediate values. One can suppose that ferromagnetic particles with l_f=20 mm are too short for fibers effective orientation and particles are too long and they are badly oriented in electromagnetic field. Therefore, the optimal length of ferromagnetic particles makes up approximately 40 mm. Values η reaching the asymptotic branch at $t \geq 180$ s supposes superposition ferromagnetic particles length-time [15].

The governing parameter of carbon plastics structure η defines it characteristics that follows from the dependence of structure fractal dimension d_f on η, adduced in Figure 3.8. As in Figures 1.2 and 1.3 for l_f=40 mm, this dependence is linear too, d_f growth at η increase is observed, but data larger number allows to obtain the more general correlation, where the limiting cases are of the greatest interest, i.e. sections, where the value d_f is independent on η and is controlled by polymeric matrix molecular and structural characteristics. As it follows from the data of Figure 3.8, at $\eta \leq 0.15$ the values d_f deviate from linear dependence and obey to the condition d_f=const=2.25. This border value d_f can be theoretically estimated by two modes [15]. Firstly, within the frameworks of thermal cluster conception the greatest value φ_{cl} is given by the equation (1.3) at the condition $\beta_T=\beta_p=0.40$ [16]. As it was shown above, this limiting value φ_{cl} is equal to 0.74. In its turn, the limiting value C_∞ can be estimated with the aid of the empirical equation (1.26), which gives C_∞=2.2. Further, the calculation according to the equation (1.4) at the indicated limiting values φ_{cl} and C_∞ gives the smallest value d_f=2.17 for carbon plastics, that well corresponds to value d_f=2.25, obtained from the plot of Figure 3.8. The second mode of d_f lower boundary estimation supposes the equation (1.16) usage, according to which at C_∞=2.20 d_f=2.17 again. Therefore, theoretical estimations of lower border value d_f (shaded horizontal line 1 in Figure 3.8) correspond well to an obtained one from the plot of Figure 3.8.

The upper limiting value d_f can be estimated by a similar mode, but for this in the equation (1.3) the exponent 1.60 [16] is used and C_∞=3. Then the smallest value φ_{cl}=0.286 and according to the equation (1.4) the greatest value d_f is equal to 2.56, that corresponds well to the data of Figure 3.8 again. Hence, the dependence $d_f(\eta)$ for carbon plastics in the general case can be expressed as follows [15]:

$$d_f = 2.17, \text{for} \eta \leq 0.15,$$
$$d_f = 2 + 1.77\eta, \text{for} 0.15 < \eta < 0.35, \qquad (3.11)$$
$$d_f = 2.56, \text{for} \eta \geq 0.35.$$

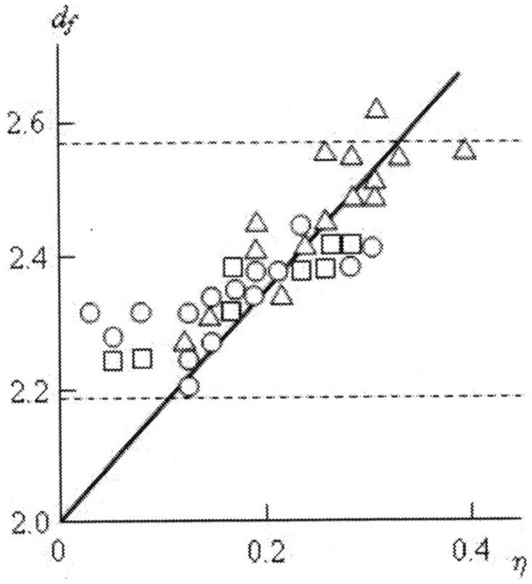

Figure 3.8. The dependence of structure fractal dimension d_f on fibers orientation factor η for carbon plastics on the basis of phenylone. The horizontal shaded lines indicate theoretical smallest (1) and greatest (2) values d_f. The designations are the same as in Figure 3.7 [15].

One can expect, that structural characteristic d_f defines carbon plastics properties, that allows to obtain relationship structure-properties for them. For this postulate verification the authors [15] choose elasticity modulus E_c, characterizing carbon plastics stiffness. In Figure 3.9 the dependence $E_c(d_f)$ is adduced, which turned out to be common for all three l_f values. Let us estimate the limiting magnitudes E_c, proceeding from the determined above corresponding d_f values. As it is known [7], dimensions d_f and D_f are connected with each other by the equation (1.22). Further, using the equation (3.9) at σ_Y=const=230 MPa and the calculated above limiting d_f values, let us obtain the smallest value E_c=1.35 GPa and the greatest one – E_c=3.21 GPa, that excellently corresponds to the data of Figure 3.9. The plot $E_c(d_f)$ extrapolation to d_f=2.0 gives $E_c \approx 1.1$ GPa and E_c calculation according to the equation (3.9) at $D_f = d_f = 2$ gives the value $E_c \approx 1$ GPa, i.e. close enough magnitudes.

Hence, the stated above results showed that fibers orientation factor defines carbon plastics structure formation, but only within the definite limits, controlled by polymeric matrix molecular and structural characteristics. The fractal analysis application allows to estimate precisely enough structural characteristics possible variation and proceeding from the latter – to calculate mechanical properties variation. The optimal length of nonequiaxial ferromagnetic particles, applied at components blending, exists, which allows to obtain the greatest fibers orientation.

Let us consider further elastic properties for polymer composites with variable contents of short fibers φ_f. As it is known [3, 17], mechanical properties of polymer composites, filled with short fibers, are defined to a considerable extent by these fibers length. The formula, derived for composites strength calculation in assumption of stresses homogeneous field along fiber, should be modified for accounting for two limiting cases. In the first of these cases fibers have length l larger than critical length l_c and can be loaded up to failure. In this case the composite strength σ_f^c is calculated according to the formula [17]:

$$\sigma_f^c = \sigma_f \varphi_f \left(1 - l_c / 2l\right) + \sigma_f^m \left(1 - \varphi_f\right), \qquad (3.12)$$

where σ_f is filler fiber strength, φ_f is fibers volume contents, σ_f^m is polymeric matrix strength.

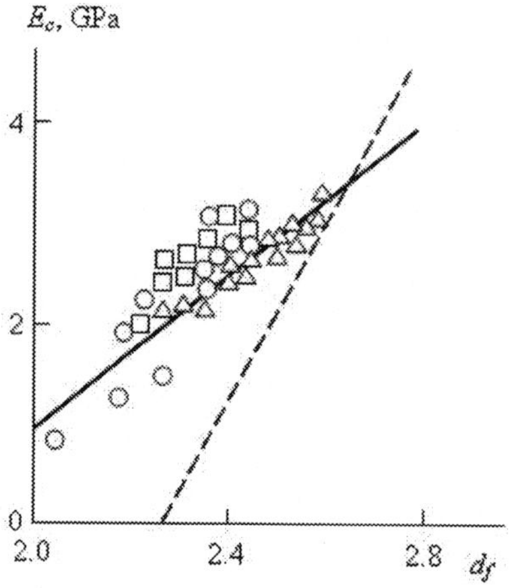

Figure 3.9. The dependence of elasticity modulus E_c on structure fractal dimension d_f for carbon plastics on the basis of phenylone. The designations are the same as in Figure 3.7 [15].

In the second case fiber length is smaller than critical one ($l<l_c$). One meets this condition in practice more rarely, but one should take into account fibers probable fracture in preparation process, decreasing their initial length [18]. This situation meets in practice often enough, since the used for polymers filling high-modulus fibers are brittle enough. Therefore the authors [19] studied the factors, influencing on elasticity modulus value of composites on the basis of polyarylate (PAr), filled with different types of short fibers.

In Figure 3.10 the dependence of failure stress σ_f^c on fiber average length $\bar{l}$ for composites on the basis of PAr, filled with uglen, vniivlon and glassy fiber, is adduced. As one can see, at small $\bar{l}$ sharp enough σ_f^c raising is observed and then this dependence reaches plateau and some σ_f^c reduction at $\bar{l}$ growth is even observed even.

Let us use the equation (3.12) and the experimental data for composites PAr-glassy fiber with the greatest value $\bar{l}$=360 mcm for estimation fiber critical length l_c. Then l_c is equal to 252 mcm. This value l_c points out in Figure 3.10 by vertical shaded line. As one can see, this line actually divides the dependence $\sigma_f^c(\bar{l})$ into two sections mentioned above, that confirms its calculation correctness.

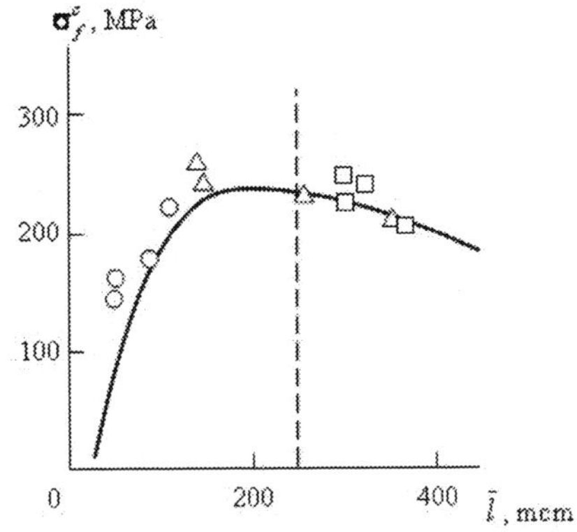

Figure 3.10. The dependence of failure stress σ_f^c on fiber average length $\bar{l}$ for composites PAr-uglen (1), PAr-glassy fiber (2) and PAr-vniivlon (3). The vertical shaded line points out the fiber critical length l_c [19].

For composites, filled with fibers with $\bar{l}<l_c$, the value σ_f^c can be calculated according to the equation [17]:

$$\sigma_f^c = \frac{\tau_i \bar{l}}{\bar{d}}\varphi_f + \sigma_f^m(1-\varphi_f), \tag{3.13}$$

where $\bar{d}$ is fiber average diameter, τ_i is shearing breaking stress on division boundary fiber-matrix or on matrix depending on the fact, which of these values is smaller.

The interfacial layer failure stress σ_a can be estimated according to the equation [20]:

$$\sigma_a = \sqrt{3}\tau_i \,. \tag{3.14}$$

The estimations according to the equations (3.13) and (3.14) have shown that for the studied composites with $\bar{l}<l_c$ the value σ_a varies within the limits of 26-146 MPa. Since failure stress for the initial PAr is equal to 168 MPa, then this means, that in composites deformation process interfacial layer is subjected to failure.

In Figure 3.11 the dependence of σ_a on interfacial layer relative fraction φ_{if} for six composites PAr-uglen and PAr-glassy fiber is shown. As one can see, the linear σ_a raise at φ_{if} growth is observed, i.e. interfacial regions relative fraction increases their strength. Let us note, that for particulate-filled composites PHE-Gr the opposite tendency was obtained - σ_a reduction at φ_{if} growth [21]. As it was noted above, such distinction is explained by different mechanisms of interfacial regions formation, which in its turn are due to different structure of filler surface. For fibers uglen and glassy fiber with relatively smooth surface and, consequently, small values of its fractal dimension d_{surf}, which are close to 2.0, polymer macromolecules are stretched on this surface, forming dense and strong enough interfacial layer [22]. In composites PHE-Gr at large enough φ_f filler particles aggregation is observed [23], owing to the value d_{surf} for these aggregates is large enough (~ 2.70) and macromolecules in interfacial layer preserve statistical coil conformation [22], as a result the interfacial regions structure in these composites is loose enough and their strength is small (~ 6-23 MPa) [21].

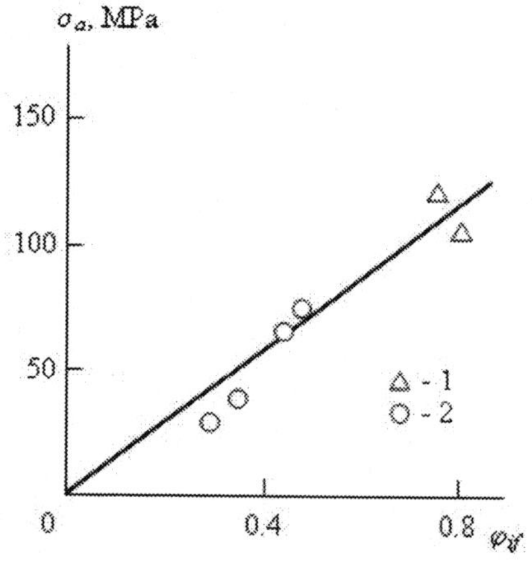

Figure 3.11. The dependence of interfacial layers strength σ_a on their relative fraction φ_{if} for composites PAr-uglen (1) and PAr-glassy fiber (2) [19].

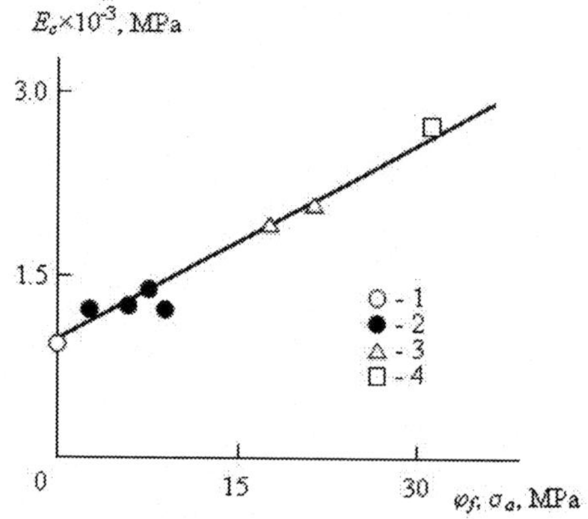

Figure 3.12. The dependence of elasticity modulus E_c on product $\varphi_f\sigma_a$ value for initial PAr (1) and composites PAr-uglen (2), PAr-glassy fiber (3) and PAr-vniivlon (4) [19].

One should expect [3, 17], that σ_a increasing results to composites elasticity modulus E_c raising. For this supposition confirmation the authors [19] plotted the dependence of elasticity modulus experimentally determined value E_c on product $\varphi_f\sigma_a$, which is shown in Figure 3.12. As one can see, in this case the good linear correlation is obtained, which is analytically expressed as follows [19]:

$$E_c = 900 + 56\varphi_f \sigma_a \text{, MPa.} \quad (3.15)$$

The dependence $E_c(\varphi_f\sigma_a)$ at φ_f=0 or σ_a=0 is extrapolated to elasticity modulus value for initial PAr, which is equal to 900 MPa, that has been expected in virtue of the obvious condition φ_f=0. Therefore, the equation (3.15) shows that the value E_c depends on the initial polymer elasticity modulus, filler volume fraction and interfacial regions strength, but is independent on fibers elasticity modulus E_f, which changes for the studied composites within the wide enough limits (~ 15-114 GPa [24]).

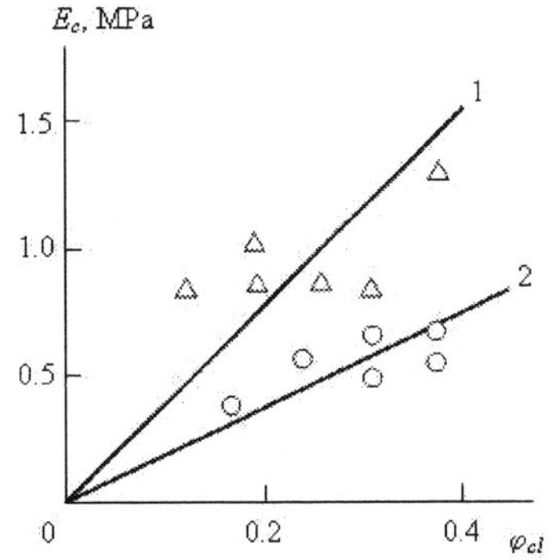

Figure 3.13. The dependences of elasticity modulus E_c on clusters relative fraction φ_{cl} for carbon plastics on the basis of HDPE at testing temperatures 293 (1) and 313 K (2) [25].

In paper [25] the elasticity modulus behaviour of composites was studied on the basis of HDPE, filled with short carbon fibers with volume contents φ_f within the range 0.038-0.303. At present elasticity modulus E and structural characteristics of semicrystalline polymers, to which HDPE belongs, are well known. So, the authors of paper [26] assume E growth at polymer crystallinity degree K increasing and in paper [27] the linear dependence of E on local order domains (clusters) relative fraction φ_{cl} for HDPE series was obtained. For these conceptions verification in case of composites with semicrystalline matrix the calculation of structural parameters K and φ_{cl} for the studied carbon plastics was fulfilled as follows. The experimental estimation of crystallinity degree, which is crystalline phase integral characteristics, can be fulfilled in assumption of polymeric matrix and filler densities additivity according to the equation (1.82), accepting carbon fibers density equal to 1320 kg/m^3 [24]. Further the mass crystallinity degree K was calculated according to the known formula (1.83). The value φ_{cl} can be calculated from the combination of the equations (1.1), (1.2), (1.4) and (1.16).

In Figure 3.13 the dependence E_c on φ_{cl} for carbon plastics on the basis of HDPE at two testing temperatures is adduced. As one can see, the expected linear correlation $E_c(\varphi_{cl})$, passing through coordinates origin, is obtained. However, the obvious lacks of this correlation are large enough data scatter (particularly at T=293 K) and two linear dependences availability – for each of the used testing temperatures T. The correlation $E_c(K)$ has a similar form.

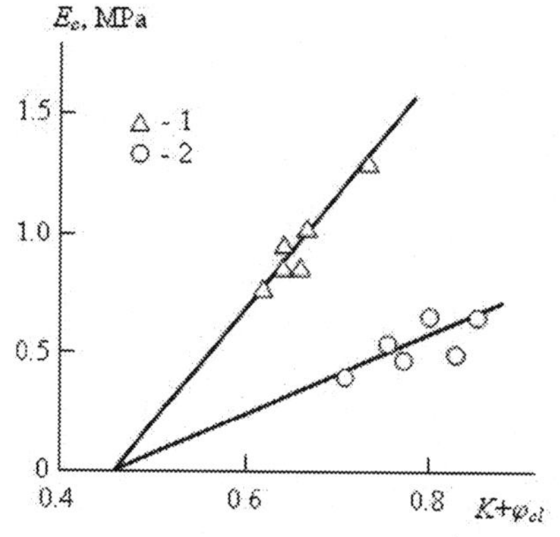

Figure 3.14. The dependences of elasticity modulus E_c on the sum relative fraction of long-order and short-order regions (K+φ_{cl}) for carbon plastics on the basis of HDPE at testing temperatures 293 (1) and 313 K (2) [25].

In Figure 3.14 the dependence of E_c on the sum of relative fractions of long-order and short-order (local) regions (K+φ_{cl}) is adduced. The two linear correlations for testing temperatures 293 and 313 K are obtained again, but the data scatter is essentially smaller. Besides, both linear dependences $E_c(K+\varphi_{cl})$ of Fig, 3.14 are extrapolated to E_c=0 at nonzero value (K+φ_{cl}), which is equal to ~ 0.45. It is obvious, that the last value is a structural parameter, dividing semicrystalline polymer and rubbers, since in the scale of Figure 3.14 the value E_c for rubbers is equal to zero (usually it has an order of 1 MPa [28]).

For obtaining the single dependence of E_c on carbon plastics polymeric matrix structural characteristics at different testing temperatures, the authors [25] studied E_c dependence on the excess energy localization regions dimension D_f. As it is known [7], this dimension characterizes critical strain regions and has nonwhole value that reflects energetic excitations fractal structure [7]. The value D_f can be determined according to the equation (1.22). In Figure 3.15 the dependence $E_c(D_f)$ for the studied carbon plastics is adduced, which turns out to be linear, passing through coordinates origin and is approximated by the single correlation for both testing temperatures, expressed analytically as follows [25]:

$$E_c = 0.19D_f \text{, GPa.} \tag{3.16}$$

Therefore, the dimension D_f is a more general parameter, characterizing polymer composites structure than K, φ_{cl} or their sum. Let us remind that this dimension characterizes polymer structure linear scales hierarchy and is equal approximately to their main molecular characteristic C_∞ [29].

Let us note an interesting feature of equations similar to the relationship (3.16). From the plot of Figure 3.4 one can see, that for phenylone the constant coefficient is equal to 0.69 in comparison with 0.19 for HDPE. Such relation supposes proportionality of the constant coefficient before D_f to matrix polymer elasticity modulus E_m and coefficient itself can be presented as the ratio E_m/E_{max}, where E_{max} is the greatest value of elasticity modulus for polymers. E_{max} estimation gives magnitude of 4 GPa order, that according to the reference data is close actually to the upper boundary for E_m.

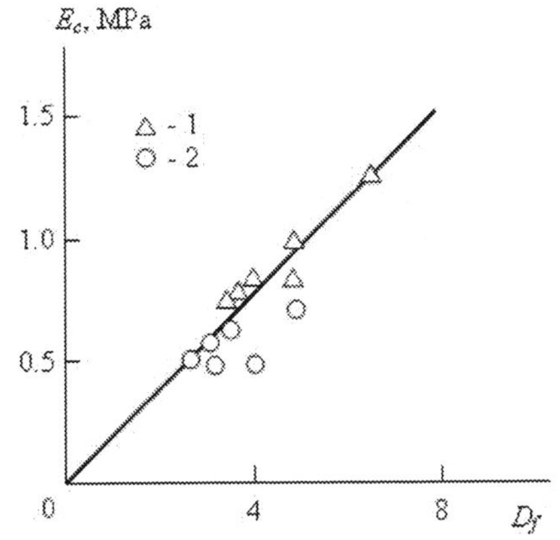

Figure 3.15. The dependence of elasticity modulus E_c on the excess energy localization regions dimension D_f for carbon plastics on the basis of HDPE at testing temperatures 293 (1) and 313 K (2) [25].

Within the frameworks of fractal analysis the value E_c can be theoretically estimated according to the equation (3.6). In Figure 3.16 the comparison of the obtained experimentally and calculated according to the indicated equation E_c values for studied carbon plastics is adduced, from which theory and experiment good correspondence follows. As it was noted above, the equation (3.6) correctness supposes insignificant role of fluctuation free volume in elastic deformation process, since mechanical tests of carbon plastics on the basis of HDPE are fulfilled by a tension scheme. Let us remind, that the equation (3.6) was obtained for particulate-filled composites PHE-Gr with amorphous matrix. The obtained correspondence of the experimental and calculated according to this equation E_c values for composites with semicrystalline matrix, filled with short fibers, speaks about high enough community degree of the given equation.

Hence, the adduced above results showed that the obtained earlier for semicrystalline polymers dependences of elasticity modulus on the structure local and long-order degree are also true for composites on the basis of such polymers. However, the obtained within the frameworks of fractal analysis dependence of elasticity modulus on the excess energy localization regions dimension or polymeric matrix structural hierarchy possesses a larger community. This circumstance confirms conception of the dependence of elasticity modulus of composites on polymeric matrix structure, modified by filler introduction.

As it was noted above, one of the main tasks, deciding at polymers filling, is the latter stiffness raising, which is characterized by elasticity modulus E_c value. How successfully this task is solved can be judged by modulus efficiency coefficient k_e change, which is determined according to the equation (1.78).

The study of polymeric matrix structure influence on the value k_e of carbon plastics on the basis of phenylone is of particular interest. These materials were prepared by technology of components preliminary blending in rotating electromagnetic field using the same binding and filler at constant φ_f value. Therefore, in the equation (1.78) the following conditions are used: E_m=const, φ_f=const and E_f=const. This means, that the only factor, influencing on the value E_c and, hence, k_e is polymeric matrix structure. Proceeding from this, the authors [31] undertook the study of concrete structural parameters, defining filling efficiency of carbon plastics on the basis of phenylone.

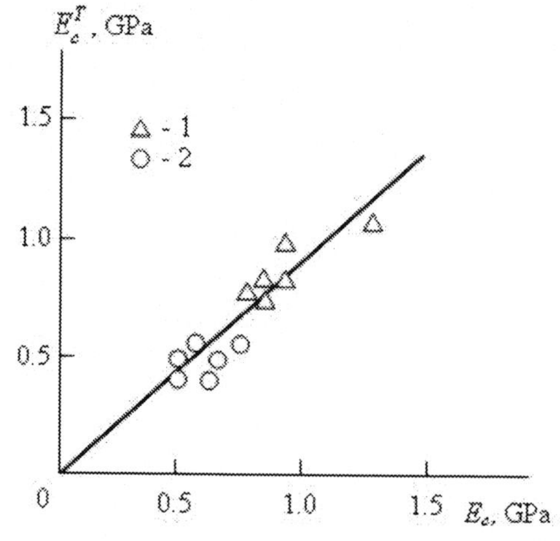

Figure 3.16. The comparison of the obtained experimentally E_c and calculated according to the equation (3.6) E_c^T elasticity modulus value for carbon plastics on the basis of HDPE at testing temperatures 293 (1) and 313 K (2) [25].

As it was shown in section 1.1, the structure of carbon plastics on the basis of phenylone is a synergetic system, the governing parameter of which serves carbon orientation factor η [10]. The estimations k_e according to the equation (1.78) showed this parameter change as a function of components blending duration in rotating electromagnetic field t within the range of 0.01-0.78. What is more, at ferromagnetic particles of length l_f=20 mm and small t (5-20 s) using the negative values k_e=-0.25÷-0.67 were obtained. The physical significance of k_e negative values is obvious: these values were obtained for those carbon plastics, which have $E_c<E_m(1-\varphi_f)$. In other words, in this case carbon fibers introduction decreases composites elasticity modulus in comparison with the initial matrix polymer (phenylone) modulus. This question will be considered lower in detail. It was found out, that the dependence $k_e(t)$ has synergetic character, similar to the function $\eta(t)$ behaviour [10]. This gives grounds to suppose certain interconnection of k_e and η values. In Figure 3.17 the dependence k_e on $\eta^{1/2}$ for four series of the studied carbon plastics is adduced. Such dependence form was chosen with the purpose of its linearization. As it follows from the data of Figure 3.17, the function $k_e(\eta^{1/2})$ is really linear in both positive and negative values k_e region and is analytically described by the following equation [31]:

$$k_e = 3.2\left(\eta^{1/2} - 0.37\right). \qquad (3.17)$$

From the equation (3.17) it follows, that the increasing Ec in comparison with Em is observed at the condition [31]:

$$\eta > 0.14. \qquad (3.18)$$

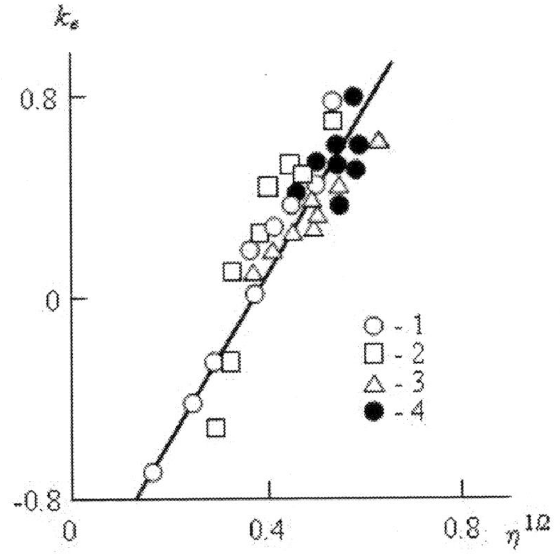

Figure 3.17. The dependence of modulus efficiency coefficient k_e on fibers orientation factor η for carbon plastics on the basis of phenylone. At components blending in rotating electromagnetic field ferromagnetic particles of length 20 (1, 2) and 40 mm (3, 4) were used with magnetic (1, 3) and mechanical (2, 4) separation application [31].

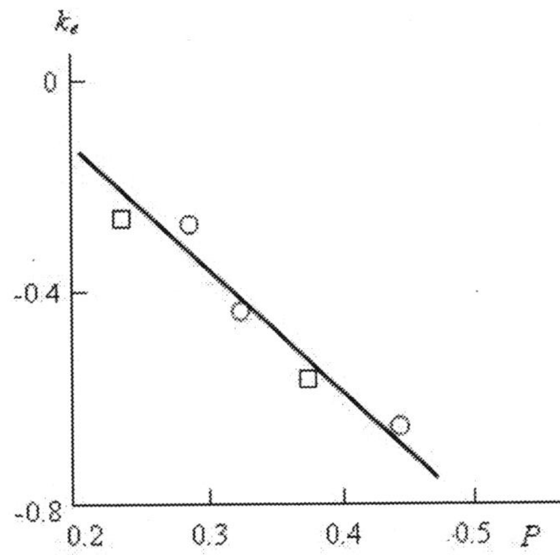

Figure 3.18. The dependence of modulus efficiency coefficient k_e on structure relative porosity P for carbon plastics on the basis of phenylone. The designations are the same as in Figure 3.17 [31].

The negative value k_e, following from the condition $E_c<E_m(1-\varphi_f)$, is defined by polymeric matrix structure porosity, which is due to filler introduction at insufficient for components complete blending values t. The value of relative porosity P can be calculated according to the following equation [4]:

$$E_c = E_m\left(1+11\varphi_f^{t_p}\right)\left(1-\frac{P}{0.84}\right)^{t_p}, \tag{3.19}$$

where tp is percolation index, equal to 1.7.

In Figure 3.18 the dependence $k_e(P)$ for five carbon plastics with negative values k_e is adduced. And as one should expect, k_e reduction at P growth is observed, that results to expected composite elasticity modulus decrease at its structure porosity raising.

To achieve high enough efficiency of filling a good connection polymer-filler is necessary [32], which is characterized quantitatively by interfacial layer strength σ_a [21]. In Figure 3.19 the dependence $k_e(\sigma_a)$ for carbon plastics, prepared with ferromagnetic particles of length 40 mm using, is adduced. As one can see, the data of this Figure confirm the supposition about filling efficiency increasing at σ_a growth. The large scatter of data shows that in this case it is more correct to speak about tendency, but not about correlation between k_e and σ_a.

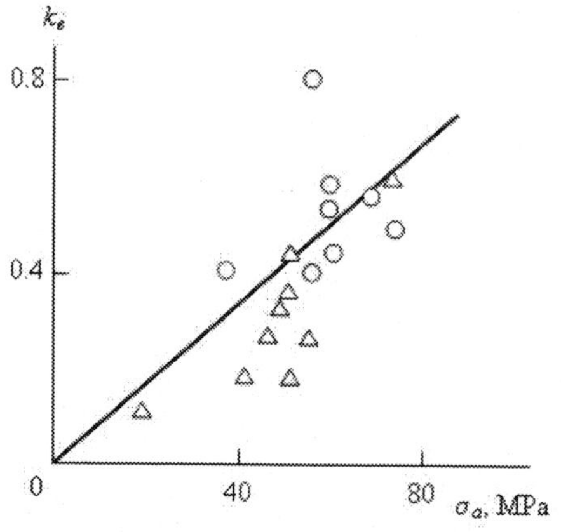

Figure 3.19. The dependence of modulus efficiency coefficient k_e on interfacial layer strength σ_a for carbon plastics on the basis of phenylone. The designations are the same as in Figure 3.17 [31].

At the first sight it occurs, that fibers orientation factor η characterizes not polymeric matrix structure, but fibers distribution in it. However, as it is known [8], in virtue of its definition as structure governing parameter the value η controls its state. So, the carbon plastics main structural characteristic, structure fractal dimension d_f, linearly increases at η growth [15]. In Figure 3.8 the dependence $d_f(\eta)$ for four series of the studied carbon plastics is adduced, which was shown d_f linear growth at η increasing. At η=0 dimension d_f=2.0.

However, as it showed in paper [33], this condition is unattainable for real polymers – a local order degree φ_{cl} increase, defining d_f reduction according to the equation (1.4), is possible only up to the definite level and further φ_{cl} increase is balanced by entropic tightness of chain parts between clusters and φ_{cl} growth ceases. This structure state was called as quasiequilibrium one [33]. Corresponding to it dimension d_f^{qe} is shown in Figure 3.8 by horizontal shaded line 1. It is significant that carbon plastics with negative values k_e have the smallest values d_f within the range of 2.21-2.29, i.e. for them the condition $d_f \approx d_f^{qe}$ is realized. This effect is the consequence of short ferromagnetic particles of length 20 mm and small t ($\leq$20 s) usage. Strictly speaking, orientation factor η is a governing parameter for interfacial regions [8] and at small η the small values φ_{if} are observed. In virtue of usual for synergetic systems feedback effect the values φ_{cl} and φ_{if} are connected with each other by the relationship (1.10). φ_{if} decrease at small t results to φ_{cl} growth according to the relationship (1.10) and d_f reduction according to the equation (1.4). However, in the conditions of polymeric matrix structure quasiequilibrium state chains between clusters are completely stretched and further φ_{cl} growth results to the mentioned matrix structure "pulling", i.e. to pores formation. This factor gives negative k_e values and results in the long run to the condition $E_c<E_m(1-\varphi_f)$ [4]. Let us note, that from the dependence $d_f(\eta)$ the reaching of d_f^{qe} =2.17 at about critical value η=0.15 (the equation (3.11)) follows.

Hence, the results stated above have demonstrated that a the only factor, influencing on efficiency of phenylone filling by short carbon fibers, is polymeric matrix structure, characterized by its governing parameter – fibers orientation factor. In its turn, the last factor defines unequivocally the polymeric matrix main structural characteristic – its fractal (Hausdorff) dimension (see the equation (3.11)). The important role in carbon plastics filling efficiency raising interfacial layers polymer-filler strength plays.

In paper [34] the structural aspects of filling efficiency for carbon plastics on the basis of HDPE, filled with carbon fibers (CF) with volume contents φ_f=0.038-0.303 were considered. The value k_e estimation according to the equation (1.78) at E_m=800 MPa and E_f=15 GPa shows very sharp reduction of this parameter at φ_f increasing within the range of 0.038-0.303 for the studied carbon plastics HDPE-CF (Figure 3.20). Let us note, that substantially higher values k_e were obtained for particulate-filled composites PHE-Gr (~ 0.4-1.7) [35] and comparable (~ 0.15-0.23) – for composites polyamide-6/Kevlar [36]. As it is known, one of the main parameters of semicrystalline polymers, defining to a great extent their properties, is crystallinity degree K (see Figure 3.14). The value K calculation for carbon plastics HDPE-CF was fulfilled according to their density measurements, this parameter at φ_f variation additive change was supposed at that (see the equations (1.82) and (1.83)). These estimations demonstrated K essential reduction at φ_f raising – from 0.72 for the initial HDPE up to 0.33 for the composite HDPE-CF with φ_f=0.303. In Figure 3.21 the dependence $k_e(K)$ is shown, from which linear growth k_e at K increasing follows, two linear sections with sharply differing slopes are observed at that. At small crystallinity degree ($K<0.47$) k_e the increase is small and this parameter absolute values are also small (k_e=0.053-0.072). At large enough values K (>0.50) the sharp growth k_e is observed and at K=0.59 the value k_e is high enough (k_e=0.84). It is interesting to note, that border K value for these sections corresponds approximately to crystallinity degree, dividing low and high density polyethylenes [30]. Consequently, the adduced data allow to suppose that the experimentally observed reduction E_c from 1250 up to 800 MPa within the range of φ_f=0.038-0.303 for carbon plastics on the basis of HDPE is due to crystallinity degree reduction namely, caused by filler introduction. In other words, realization of high-modulus composites on the basis of semicrystalline polymers requires high-crystalline matrix. So, the preservation of the initial HDPE value K=0.70 for composite HDPE-CF with φ_f=0.303 gives $E_c\approx$7600 MPa according to the equation (1.78), that is on the order of magnitude larger than experimentally observed value E_c (~ 800 MPa) for this carbon plastic [34].

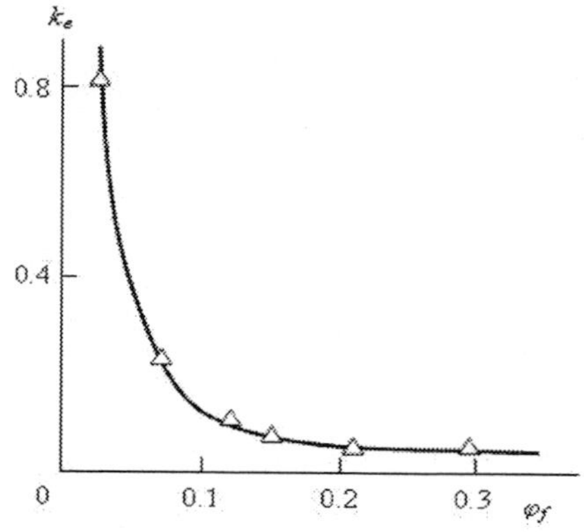

Figure 3.20. The dependence of modulus efficiency coefficient k_e on filler volume contents φ_f for composites HDPE-CF [34].

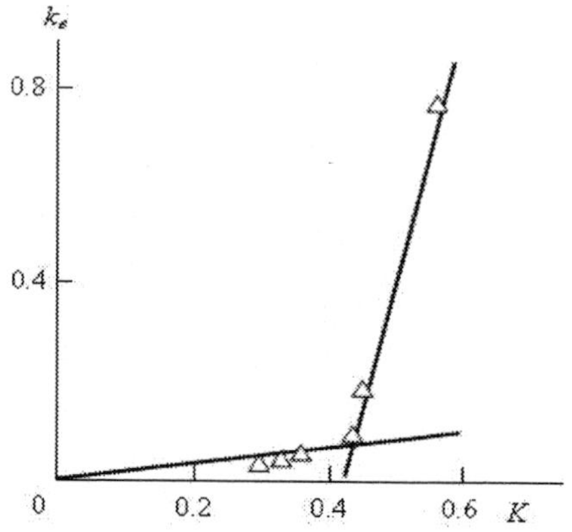

Figure 3.21. The dependence of modulus efficiency coefficient k_e on crystallinity degree K for composites HDPE-CF [34].

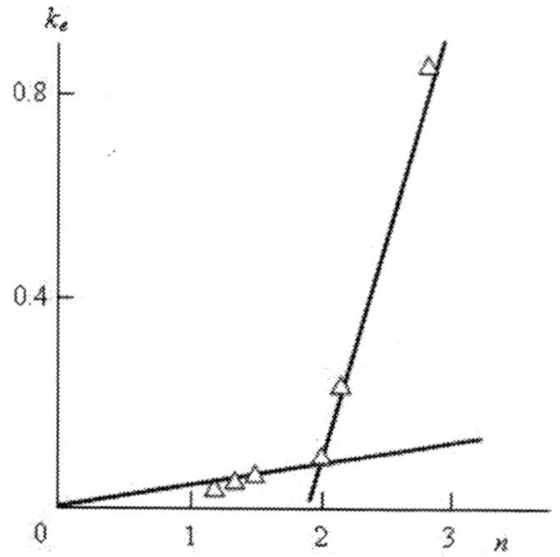

Figure 3.22. The dependence of modulus efficiency coefficient k_e on the exponent n in Kolmogorov-Avrami equation for composites HDPE-CF [34].

The value k_e depends not only on crystalline phase integral characteristic K, but also on its morphology, which can be characterized by Kolmogorov-Avrami exponent n (see the equation (1.91)). In paper [37] it has been shown, that the exponent n depends on fractal dimension D_{ch} of chain part between clusters in HDPE amorphous phase and this dependence is given by the equation (1.92).

In Figure 3.22 the dependence of k_e on the exponent n value is adduced, which is similar by shape to the correlation $k_e(K)$, shown in Figure 3.21. At small n (n<2.0) the weak k_e increase at small absolute values of this parameter (k_e=0.053-0.072) is observed and at n>2.0 the value k_e increase more than on order at n raising within the range of 2.14-2.86. As it is known [38], n variation defines the forming crystalline phase morphology of polymers. In case of athermic nucleation at $n\leq2.0$ a ribbons are formed by two-dimensional growth mechanism, at $n\leq3.0$ – circles and at n>3.0 – spheres. The fractional values n mean thermal/athermic nucleation combined mechanism, fractional part decreasing at that points out to athermic mechanism role raising. Therefore, from the data of Figure 3.22 it follows, that isotropic morphology (circles and spheres) is a preferable one and anisotropic ribbon morphology owing to exponent n reduction sharply decreases the value k_e and hence, E_c. It is obvious, that in general case introduction of filler with smooth surface results to crystals growth change from three-dimensional to two-dimensional one with corresponding exponent n reduction [38]. It is easy to trace this law by the following simple mode. The total area of carbon fibers surface will grow proportionally to $\varphi_f^{2/3}$ and adduced in Figure 1.76 the dependence $n(\varphi_f^{2/3})$ shows actually linear reduction n at filler surface total area growth. It is probable, that at φ_f>0.25 the fibers aggregation process begins, that decreases surface total area and decelerates n reduction. At $\varphi_f\approx0.55$ n value reaches 1.0, i.e. at larger φ_f athermic nucleation is realized, in other words, simultaneous beginning of all crystallites growth, is obvious, on fibers surface [38]. At φ_f=0, i.e. for initial HDPE, n=4 and in this case spherical crystalline form growth at thermal nucleation occurs, i.e. a new crystallites nucleation in crystallization process occurs [38]. The similar effect was observed for system polyamide-6/quartz powder, where crystalline phase parameters are linearly reduced at filler specific surface growth [39].

Hence, the data considered above showed that the modulus efficiency coefficient for composites with semicrystalline matrix is controlled by both crystallinity degree integral value and crystalline phase morphology, i.e. by polymeric matrix structure again. Fibers surface total area increasing results to transition from three-dimensional crystallization to two-dimensional one and to nucleation type change from thermal to athermic one. In the long run this reduces both crystallinity degree and elasticity modulus of composites.

3.2. THE YIELD PROCESS OF POLYMER COMPOSITES

For engineering polymer materials yield reaching means further impossibility of this material usage and in virtue of this circumstance yield stress σ_Y can be considered as its exploitation upper boundary. Therefore the yield process is an object of intensive study and for its description a large number of conceptions was proposed. Particularly, at present proportionality of σ_Y and elasticity modulus E are supposed for polymeric materials. Polymer composites on the basis of phenylone, filled with short carbon fibers, discovered two interesting features of yield process [40]. Firstly, in compression tests these composites showed high values σ_Y (~ 230 MPa) at relatively small E (~ 2.13-3.33 GPa) and secondly the indicated value σ_Y was approximately constant at the mentioned modulus of elasticity variation. Therefore the authors [41] fulfilled quantitative description of yield stress such behaviour for the indicated composites within the frameworks of two conceptions: thermodynamic (anharmonicity theory [13]) and structural (cluster model of polymers amorphous state structure [42]) ones.

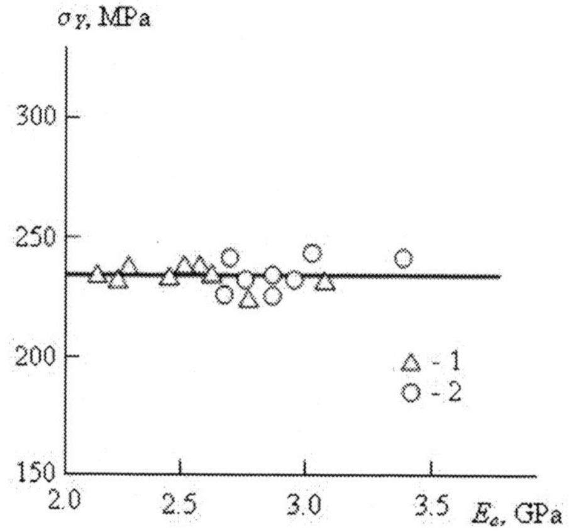

Figure 3.23. The dependence of yield stress σ_Y on elasticity modulus E_c for composites on the basis of phenylone, prepared with magnetic (1) and mechanical (2) separation application [41].

In Figure 3.23 the dependence $\sigma_Y(E_c)$ for two series of the studied composites is adduced, prepared with magnetic and mechanical separation application. As it follows from the data of this Figure, σ_Y value is high enough (~ 230 MPa) and is practically independent on elasticity modulus E_c. Within the frameworks of thermodynamic anharmonicity theory σ_Y change can be described quantitatively with the aid of the equation (3.4), in which the lattice (i.e., accounting for anharmonicity of both intra- and intermolecular bonds [43]) Grüneisen parameter γ is determined according to the equation (3.5). The comparison of experimental σ_Y and calculated according to the equations (3.4) and (3.5) at A=0.74 theoretical σ_Y^T values of yield stress together with the corresponding γ values is adduced in table 3.1. As one can see, the good enough correspondence of theory and experiment (σ_Y and σ_Y^T average discrepancy makes up ~ 10.5 %, that approximately corresponds to precision of mechanical characteristics experimental determination).

Within the frameworks of the cluster model of polymers amorphous state structure yield stress is determined as follows [44]:

$$\sigma_Y = \frac{Eb\rho_d^{1/2}}{2\pi}, \qquad (3.20)$$

where b is Burgers vector, ρ_d is density of structure linear defects (analog of dislocations in crystals).

The including in the equation (3.20) parameters are determined as follows. The value b is estimated according to the equation [44]:

$$b = \left(\frac{60.7}{C_\infty} \right)^{1/2}, \text{Å}, \quad (3.21)$$

where C∞ is characteristic ratio.

Table 3.1. The yield process parameters and structural characteristics for composites on the basis of phenylone as a function of components blending duration *t* in rotating electromagnetic field [41]

t, s	σ_Y, MPa	γ	E_c, GPa	σ_Y^T, MPa, the equation (3.4)	C_∞	b, Å	σ_Y^T, MPa, the equation (3.20)
The magnetic separation							
5	222	1.92	2.72	236	5.83	3.21	224
10	235	1.20	2.13	295	3.75	4.02	232
20	235	1.29	2.25	290	5.03	3.47	245
30	234	1.48	2.46	277	5.55	3.31	245
60	228	2.24	3.02	225	5.62	3.29	215
120	232	1.55	2.61	280	5.47	3.33	247
180	235	1.58	2.58	273	5.49	3.33	245
240	230	1.31	2.23	284	4.81	3.55	243
300	230	1.51	2.45	270	5.25	3.40	237
The mechanical separation							
5	228	2.16	2.97	229	6.46	3.07	234
10	225	2.21	2.96	223	7.31	2.88	259
20	229	1.72	2.63	255	4.94	3.51	230
30	228	1.80	2.72	252	6.11	3.15	239
60	236	2.02	2.97	245	6.33	3.10	244
120	235	1.67	2.65	265	3.61	4.10	240
180	232	2.66	3.33	209	6.57	3.04	224
240	231	1.91	2.82	246	6.58	3.04	250

In the cluster model all segments, entering polymer structure (for composites-polymeric matrix structure) densely-packed regions, are considered as linear defects [44]. As it was shown above, for the studied composites such regions there are two types of: local order domains (clusters) in bulk polymeric matrix and interfacial regions.

Earlier within the frameworks of polymers structure hierarchical model [29] it has been shown that value C_∞ for interfacial layer (C_∞^{if}) is equal to $\left(C_\infty^{m}\right)^2$, where C_∞^{m} is C_∞ value for bulk polymeric matrix, i.e. C_∞^{if}=9. Since in σ_Y calculations the densely-packed regions total relative fraction (φ_{cl}+φ_{if}) is used, then the average C_∞ value for using in the equation (3.21) was determined according to the mixtures rule [41]:

$$C_\infty = \frac{\varphi_{cl} C_\infty^m + \varphi_{if} C_\infty^{if}}{\varphi_{cl} + \varphi_{if}}. \qquad (3.22)$$

And at last, for calculation simplification ρ_d value can be determined as follows [45]:

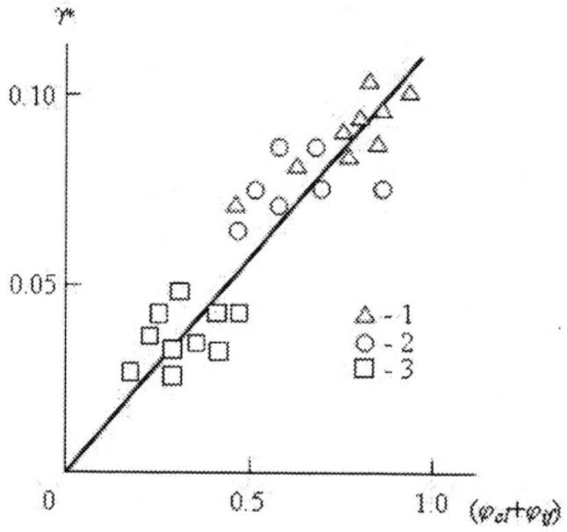

Figure 3.24. The dependence of shearing critical strain γ_* on densely-packed regions total relative fraction (φ_{cl}+φ_{if}) for carbon plastics on the basis of phenylone, prepared with magnetic (1) and mechanical (2) separation application and on φ_{cl} for epoxy polymers according to the data of paper [46] (3) [41].

$$\rho_d = \frac{\varphi_{cl} + \varphi_{if}}{S}, \qquad (3.23)$$

where S is macromolecule cross-sectional area.

In table 3.1 the results of σ_Y^T calculation according to the structural model and also obtained by the indicated mode parameters C_∞ and b are listed. As one can see, the good correspondence is obtained again between theory and experiment (the average discrepancy of σ_Y and σ_Y^T makes up ~ 4.3 %).

The composite structure stability degree at the influence of mechanical stresses on it can be estimated by shearing critical strain γ_*, at which solid loses shearing stability [46]:

$$\gamma_* = \frac{\sigma_Y}{E} = \frac{1}{6\gamma}. \qquad (3.24)$$

For the studied composites the value γ_* is relatively large and varies within the limits of 0.071-0.110. For the comparison, in the cross-linked polymers case in compression tests γ_*=0.025-0.050 [46], and for linear polymers it is much smaller and equal to ~ 0.023 [47]. Such high stability to shearing deformation for the studied composites is connected with high relative fraction of densely-packed regions ($\varphi_{cl}+\varphi_{if}\approx0.74$). This is confirmed by the data of Figure 3.24, where the dependence $\gamma_*(\varphi_{cl}+\varphi_{if})$ for the studied carbon plastics is adduced and the dependence $\gamma_*(\varphi_{cl})$ for cross-linked polymers according to the data of paper [46] is also shown. As one can see, the dependence of γ_* on densely-packed regions fraction for both composites and nonfilled cross-linked polymers is described by a common linear correlation that confirms the observed effect common origin.

As it is known [13, 43], the lattice parameter Grüneisen value characterizes intermolecular bonds weakening degree in polymers deformation process. In Figure 3.25 the dependence of γ on densely-packed regions total relative fraction ($\varphi_{cl}+\varphi_{if}$) for carbon plastics on the basis of phenylone is adduced. As it follows from the data of this Figure, ($\varphi_{cl}+\varphi_{if}$) increasing results to γ reduction. At ($\varphi_{cl}+\varphi_{if}$)=1.0 $\gamma\approx0.74$, that was expected according to the formula (3.5): at dense packing of polymer d_f=2.0 and ν=0 and then $\gamma=A$. Therefore, the larger ($\varphi_{cl}+\varphi_{if}$) smaller γ and the higher yield strain ε_Y is, which is connected with γ according to the following simple relationship [44]:

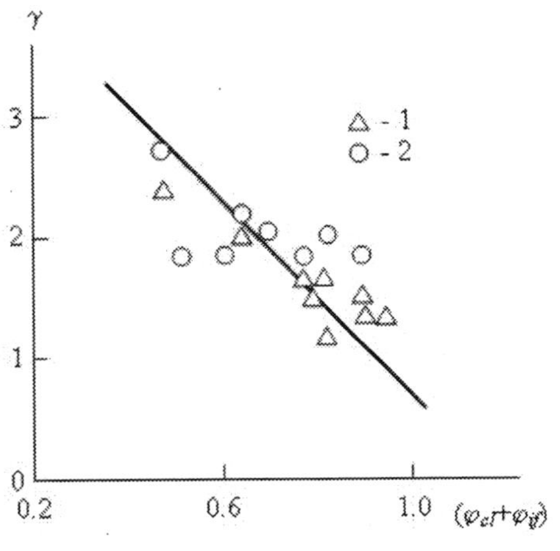

Figure 3.25. The dependence of Grüneisen parameter γ on densely-packed regions total relative fraction ($\varphi_{cl}+\varphi_{if}$) for carbon plastics on the basis of phenylone, prepared with magnetic (1) and mechanical (2) separation application [41].

$$\varepsilon_Y = \frac{1}{2\gamma}. \qquad (3.25)$$

According to the equation (3.25) it is expected that ε_Y value for the studied composites will be varied within the limits of 0.188-0.417 [41].

Hence, the stated above results demonstrated that both thermodynamic and structural treatments adequately described yield stress behaviour of carbon plastics on the basis of phenylone. The yield stress high values and structure stability to mechanical stress influence are due to densely-packed regions large relative fraction in these composites structure. Both treatments are interconnected, since Grüneisen parameter is a diminishing linear function of densely-packed regions total relative fraction.

Let us note that interconnection of yield stress and polymers structure densely-packed regions is known quite a long time. The authors [48] demonstrated experimentally, that yield process for amorphous glassy polymers was realized in their structure densely-packed regions. This question was personificated later in detail with the cluster model of polymers amorphous state structure [44, 49] and fractal analysis [29] usage. So, for linear and cross-linked amorphous polymers the yield process is controlled by local order domains (clusters) [50, 51]. In case of semicrystalline polymers besides clusters in the indicated process part of crystalline phase participates, which is subjected to partial melting-recrystallization (mechanical disordering) [45]. In polymer particulate-filled composites with amorphous glassy matrix the yield process is also controlled by clusters [52]. From practical point of view the considered question is important because that yield stress value is proportional to square root from structural regions relative fraction, controlling yield process (see the equation (3.20)). Proceeding from this, the authors [40, 53] studied yield process structural aspect again, using in this case the plasticity fractal theory.

Within the frameworks of the indicated polymers plasticity conception [54], which theoretically substantiates the postulate about yielding occurrence only in a polymer structure part, it was shown that Poisson's ratio ν_Y in yield point can be expressed by the following relationship:

$$\nu_Y = \nu\chi + 0.5(1-\chi), \tag{3.26}$$

where χ is realization probability of elastic (and $(1-\chi)$ – nonelastic one) state of different elements in deformable polymer volume. The value νY was accepted equal to 0.43 [44].

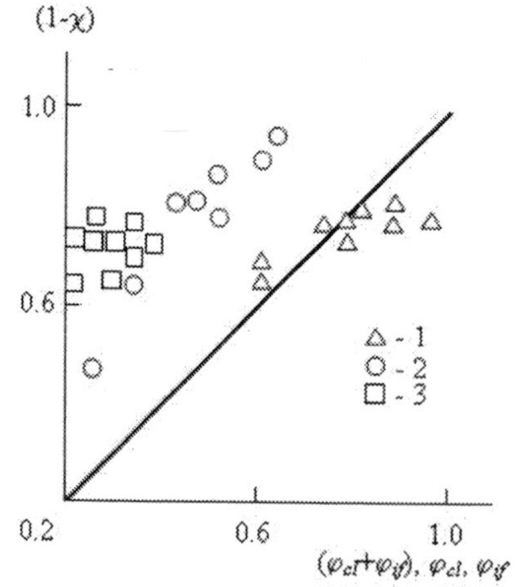

Figure 3.26. The comparison of polymeric matrix fraction $(1-\chi)$, subjected to yielding, and structural components relative fractions $(\varphi_{cl}+\varphi_{if})$ (1), φ_{cl} (2) and φ_{if} (3) for carbon plastics on the basis of phenylone. The straight line shows relation 1:1 [53].

The comparison of parameter $(1-\chi)$ and relative fractions φ_{cl} and φ_{if} (or their sum $(\varphi_{cl}+\varphi_{if})$) of structural components gives the answer to the question, which of these components controls the yield process. In Figure 3.26 such comparison for carbon plastics on the basis of phenylone is adduced. As one can see, the unequivocal correspondence between the values (1-χ) and (φ_{cl}+φ_{if}) is obtained, whereas each from separately taken densely-packed regions fractions φ_{cl} or φ_{if} is substantially smaller than (1-χ). Hence, the data of Figure 3.26 suppose that yield process in the studied carbon plastics is realized at stability simultaneous loss by both clusters and interfacial regions [44, 49]. Therefore, the stated above results confirmed conclusion the made above that clusters and interfacial regions, i.e. all densely-packed regions of amorphous glassy matrix, were polymeric matrix structural components, controlling yield process in composites filled with short fibers.

Let us consider further the yield process structural aspects in composites with semicrystalline matrix. As it is known, for similar polymers all existing conceptions take into account the crystalline phase role. In the composites case such treatment is complicated by their more complex structure in comparison with the initial matrix polymer (for example, by interfacial regions availability). Proceeding from this, the authors [55] offered the complete structural treatment of yield process for carbon plastics on the basis of HDPE within the frameworks of plasticity fractal theory.

As the estimations according to the equation (1.2) have shown, carbon fibers volume contents increase from 0.038 up to 0.303 results to Poisson's ratio ν reduction from ~ 0.375 up to 0.234, that according to the plasticity fractal conception (the equation (3.26)) reduces χ and, accordingly, the value (1-χ) increases from ~ 0.40 up to 0.70 within the indicated range φ_f. The cluster model of polymers amorphous state structure assumes [44], that yielding in polymeric materials is realized when the ordered regions with relative fraction (1-χ) lose their stability in mechanical stresses field. Let us consider carbon plastics on the basis of HDPE structure at greater length in order to understand which structural regions exactly lose their stability in yield process and thereby to define σ_Y value.

As it has been noted above, the crystallinity degree K is one of the most important characteristics of semicrystalline polymers. The value K, which serves as integral characteristic of crystalline phase, can be determined in supposition of polymeric matrix and filler densities additivity according to the equations (1.82) and (1.83). Besides the crystalline phase, the clusters in amorphous phase and interfacial regions with relative fractions φ_{cl} and φ_{if}, accordingly, are densely-packed regions in polymeric matrix structure of carbon plastics. As preliminary estimations have shown, the correspondence between parameters (1-χ) and (K+φ_{if}), the relation of which is adduced in Figure 3.27. Such correspondence indicates that at macroscopic yielding reaching of carbon plastics on the basis of HDPE crystallites and interfacial regions lose stability, whereas clusters in devitrificated amorphous phase do not influence on yield process and hence, on σ_Y value. This is a substantial distinction of carbon plastics from matrix HDPE, where both crystallites and clusters participated in yield process [45]. It can be supposed that σ_Y increase at elasticity modulus E reduction at φ_f growth is due to the fact that faster increase φ_{if} in comparison with K reduction occurs.

These estimations can be confirmed quantitatively within the frameworks of cluster model, where yield stress is determined according to the equation (3.20). In Figure 3.28 the comparison of experimental σ_Y and calculated according to the indicated equation σ_Y^T yield stress values for the studied carbon plastics is adduced. As one can see, this comparison has shown theory and experiment good correspondence. Hence, σ_Y increase at φ_f growth is defined by interfacial regions fraction increasing, compensating elasticity modulus reduction.

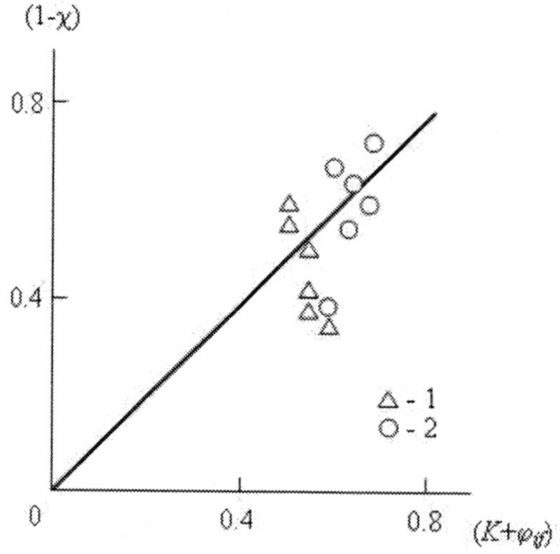

Figure 3.27. The comparison of yield process structural parameters (1-χ) and (K+φ_{if}) for carbon plastics on the basis of HDPE at testing temperatures 293 (1) and 313 K (2) [55].

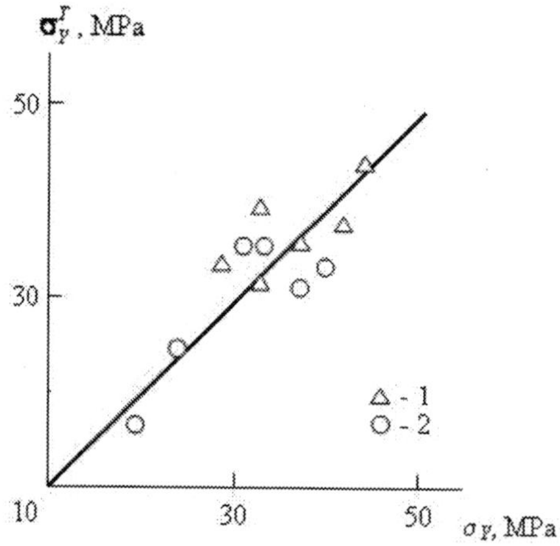

Figure 3.28. The comparison of obtained experimentally σ_Y and calculated according to the equation (3.20) σ_Y^T yield stress values for carbon plastics on the basis of HDPE at testing temperatures 293 (1) and 313 K (2) [55].

As it was noted above, carbon plastics on the basis of phenylone have two interesting features of yield process. Firstly, in compression tests these composites showed high values σ_Y (~ 230 MPa) at relatively low E_c values (~ 2.13-3.33 GPa) and, secondly, the indicated value σ_Y was approximately constant at the observed elasticity modulus variation (Figure 3.23). The explanations of these features were given above, but one should note that polymer composites in case of their usage as engineering materials can be subjected to various loading schemes: tension, compression, shear, bend and so on. In practical aspect it is important to know their behaviour at any of the indicated schemes, since this behaviour not necessarily will be identical [13]. Therefore the authors [56] fulfilled comparative structural analysis of yield process in tension and compression tests for carbon plastics on the basis of phenylone.

The yield stress value in tension tests σ_Y^{tens} can be calculated by similar stress at compression σ_Y^{comp} with the aid of the equation (3.3). In Figure 3.29 the dependences of carbon plastics yield stress in tension and compression (σ_Y^{tens} and σ_Y^{comp}, accordingly) by components blending duration in rotating electromagnetic field *t* are adduced. As one can see, these dependences character is essentially differed If in the case of compression tests yield stress is not practically changed (its variation makes up ~ 6 %, that is within the limits of experiment error), then in tension tests for samples, prepared with magnetic separation application, the clearly expressed extreme dependence is observed with σ_Y^{tens} variation more than 40 %. Hence, at carbon plastics usage according to the compression scheme *t* value does not play any essential role, whereas in the tension case σ_Y^{tens} can be increased substantially by the value *t* selection: from 129 MPa at *t*=60 s up to 175 MPa at *t*=10 s, i.e. approximately in 1.4 times. Let us note σ_Y^{tens} various behaviour for samples, prepared with magnetic and mechanical separation application, the explanation of which will be given lower. The average values σ_Y^{tens} for samples, prepared with separation indicated methods application, are equal to 154 and 137 MPa, i.e. on the average magnetic separation application gives σ_Y^{tens} approximately on 12 % higher, then in the mechanical separation case [56].

In section 1.1 it has been shown that carbon plastics structure is a synergetic system, containing two densely-packed structural components: local order domains (clusters) in bulk polymeric matrix and interfacial regions with relative fractions φ_{cl} and φ_{if}, accordingly. The values φ_{cl} and φ_{if} are connected with each other by the relationship (1.10) in virtue of feedback effect, which is specific for synergetic systems [9].

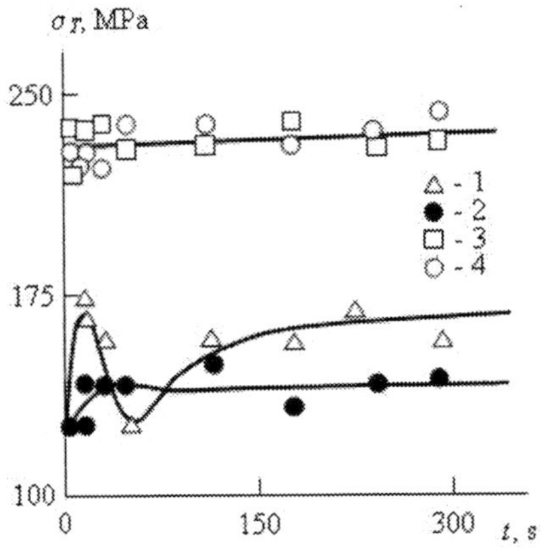

Figure 3.29. The dependences of yield stress σ_Y at tension (1, 2) and compression (3, 4) on components blending duration in rotating electromagnetic field *t* for carbon plastics on the basis of phenylone, prepared with magnetic (1, 3) and mechanical (2, 4) separation application [56].

In Figure 3.30 the dependence σ_Y^{tens} (φ_{cl}) for the studied carbon plastics is adduced, from which linear increasing σ_Y^{tens} at φ_{cl} growth follows. Let us note, that to obtain such dependence for σ_Y^{comp} is impossible, since σ_Y^{comp} =const at φ_{cl} fourfold variation. The latter is explained by the fact that in compression tests the yield stress is the function of the sum (φ_{cl}+φ_{if}) or φ_{dens}, which is approximately constant and equal to 0.74 (Figure 3.26). These results allow to suppose the following structural ground of differences in absolute values and behaviour σ_Y^{comp} and σ_Y^{tens}: if the first parameter is controlled by total value of densely-packed regions φ_{dens}, then the second – by clusters relative fraction φ_{cl} only. The conditions $\varphi_{dens}\geq\varphi_{cl}$ and $\varphi_{dens}\approx0.74$ explain the indicated difference. Besides, from the plot of Figure 3.30 it follows, that in general the values φ_{cl} for carbon plastics samples, prepared with mechanical separation application, are lower and has smaller variation in comparison with similar values for magnetic separation case. This supposes that ferromagnetic particles wear products, not removed in mechanical separation process, prevent clusters formation.

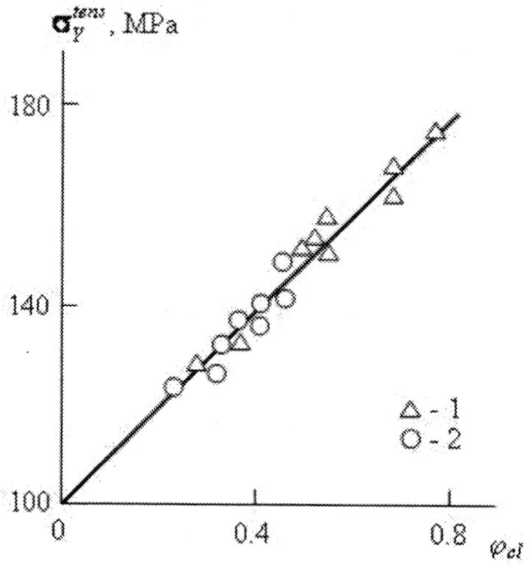

Figure 3.30. The dependence of tension yield stress σ_Y^{tens} on clusters relative fraction φ_{cl} for carbon plastics on the basis of phenylone, prepared with magnetic (1) and mechanical (2) separation application [56].

Let us consider further σ_Y and E_c relation for carbon plastics. The elasticity modulus value in the tension case can be determined according to the equation (1.2) at the condition $\sigma_Y=\sigma_Y^{tens}$. In Figures 3.23 and 3.31 the relations σ_Y/E_c for carbon plastics in case of compression and tension tests, accordingly, are adduced. As one can see, in both cases the postulated proportionality σ_Y/E [47] is not fulfilled. For compression tests the value σ_Y^{comp} is independent on elasticity modulus (Figure 3.23) and in tension case yield stress reduces in general at E_c growth. Such behaviour of the ratio σ_Y/E_c is explained within the frameworks of the cluster model of polymers amorphous state structure [44]. As it is known [44], for structure and properties of thermodynamically nonequilibrium solids description, to which amorphous polymers belong, two order parameters relate, as a minimum. The postulated in paper [47] the ratio σ_Y/E constancy supposes that the value σ_Y is defined by one order parameter only, namely, by E, that can be correct only in an individual case. In general case the value σ_Y depends on φ_{dens}, E and polymer molecular characteristics (see the equation (3.20)), owing to which the change various degree of these parameters defines the ratio σ_Y/E_c variation, as it follows from the data of Figures 3.23 and 3.31.

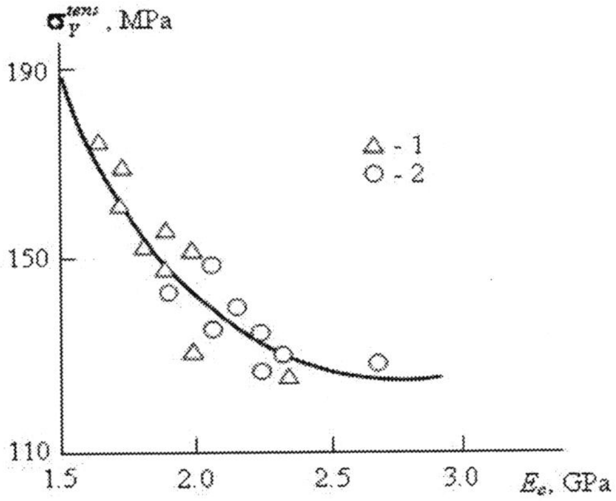

Figure 3.31. The dependence of yield stress σ_Y^{tens} on elasticity modulus E_c in tension tests for carbon plastics on the basis of phenylone. The designations are the same as in Figure 3.30 [56].

Let us consider in conclusion an energetic aspect of carbon plastics yield process. As it is known [5], the level of "pumping" in polymeric matrix energy, which is due to filler introduction, or matrix structure "disturbance" degree can be characterized by the excess energy localization regions dimension D_f, determined according to the equations (1.21) and (1.22). On the other hand, energy density W_c, "pumping" in polymer at mechanical influence, is estimated by analogy with the equation (1.52) as follows [5]:

$$W_c = \frac{\sigma_Y^2}{2E}. \qquad (3.27)$$

In Figure 3.32 the dependences $W_c(D_f)$ for the studied carbon plastics in tension and compression tests are adduced. As one can see, in both cases W_c reduction at D_f growth is observed. This means, that the higher polymeric matrix structure "disturbance" degree, due to introducing filler is, the smaller mechanical energy supplied from outside is required for yield stress reaching. By absolute value the values W_c at compression (W_c^{comp}) are higher than similar magnitudes at tension (W_c^{tens}). The dependences $W_c^{comp}(D_f)$ and $W_c^{tens}(D_f)$ are intersected at D_f=2 and the limiting value $W_c^{\lim}$=13.6 MJ/m^3. It is obvious, that the difference $W_c^{\lim}-W_c^{comp}$ (or $W_c^{\lim}-W_c^{tens}$) characterizes energy density, "pumping" in polymer at filler introduction. The value D_f=2, at which the limiting value $W_c^{\lim}$ is reached (Figure 3.32), according to the equation (1.21) corresponds to ν=0 or, according to the equation (1.1) at d=3,

d_f=2.0. As it follows from the relationship (1.4), such d_f value is reached at polymer dense packing (φ_{cl}=1.0) and the smallest value C_∞=2 at polymeric chain tetrahedral valent angles [57].

As it was noted above, structural ground of values σ_Y^{comp} and σ_Y^{tens} distinction is the fact, that at yielding in compression tests both clusters and interfacial regions lose stability, whereas in tension tests – clusters only. If this supposition is correct, then it can be written [56]:

$$\frac{W_c^{comp}}{W_c^{tens}} = \frac{\varphi_{cl} + \varphi_{if}}{\varphi_{cl}}. \tag{3.28}$$

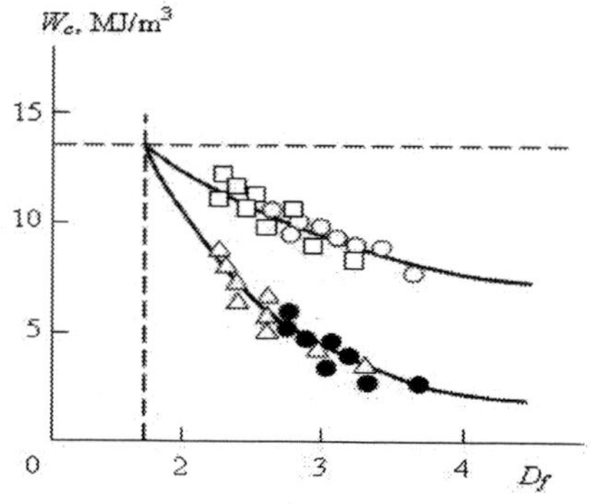

Figure 3.32. The dependences of mechanical energy density W_c on excess energy localization regions dimension D_f for carbon plastics on the basis of phenylone. The designations are the same that in Figure 3.29 [56].

In Figure 3.33 the comparison of the indicated in the equation (3.28) ratios is adduced, from which their approximate equality follows. This circumstance confirms postulated above structural ground of yield stress distinction in tension and compression tests for carbon plastics.

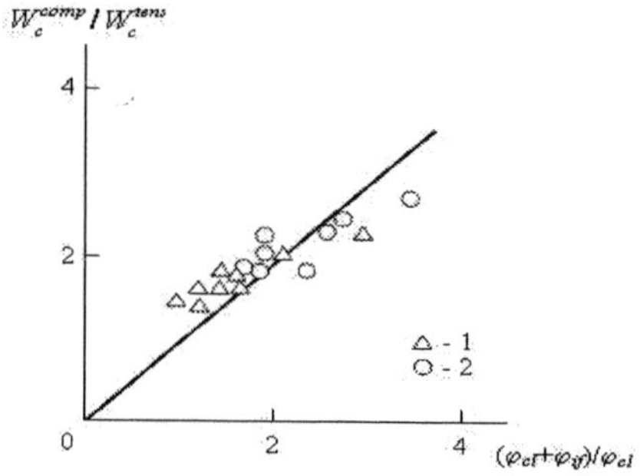

Figure 3.33. The dependences of ratio of mechanical energy densities at compression and tension W_c^{comp}/W_c^{tens} on ratio of structural components relative fractions $(\varphi_{cl}+\varphi_{if})/\varphi_{cl}$ for carbon plastics on the basis of phenylone. The straight line shows relation 1:1. The designations are the same as in Figure 3.30 [56].

Hence, the stated above results found out yield stress different dependences in tension and compression on components blending duration in rotating electromagnetic field. This difference is due to the fact, that in yield processes at the mentioned loading schemes various structural components participate. The ratio of yield stress to elasticity modulus can either remain constant or reduce [56].

3.3. COMPOSITES BEHAVIOUR ON COLD FLOW PLATEAU

Haward showed [58, 59], that polymers behaviour on cold flow plateau (forced high-elasticity) in case of homogeneous deformation could be described for compression tests by the equation:

$$\sigma^{tr} = \text{const} + G_p\left(\lambda^2 - \lambda^{-2}\right), \qquad (3.29)$$

where σ^{tr} is true stress of cold flow plateau at draw ratio λ, G_p is the so-called strain hardening modulus.

The authors [60] proposed to use the parameter G_p for estimation of physical entanglements cluster network density ν_{cl}, where clusters consisted of several densely-packed collinear statistical segments of polymeric chain, appling for this purpose the equation (1.87). The equation (3.29) describes well amorphous glassy and semicrystalline polymers on cold flow plateau [58, 59]. The authors [61] used the indicated model for description of cold flow plateau of curve stress-strain of carbon plastics on the basis of phenylone in compression tests.

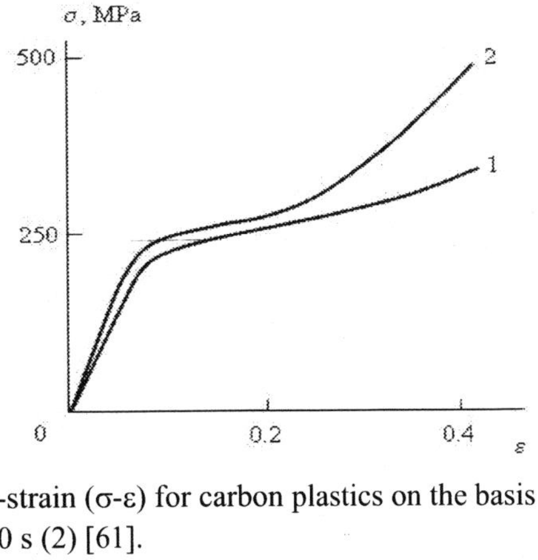

Figure 3.34. The curves stress-strain (σ-ε) for carbon plastics on the basis of phenylone at l_f=40 mm, t=30 s (1) and l_f=40 mm, t=240 s (2) [61].

In Figure 3.34 two typical curves stress-strain (σ-ε), obtained in compression tests for carbon plastics on the basis of phenylone were shown. As one can see, these curves demonstrate very strong strain hardening for carbon plastics – within the small range of draw ration λ (λ=1+ε) 1.2-1.4 the value σ increases from ~ 240 up to 340 MPa. In Figure 3.35 the dependence σ^{tr} on (λ^2-λ^{-2}), corresponding to the equation (3.29), for two carbon plastics, whose curves σ-ε are shown in Figure 3.34, are adduced. The data of Figure 3.35 show, that in both cases the linear dependences are obtained, allowing G_p calculation by their slopes, and those slopes substantially differ by absolute value.

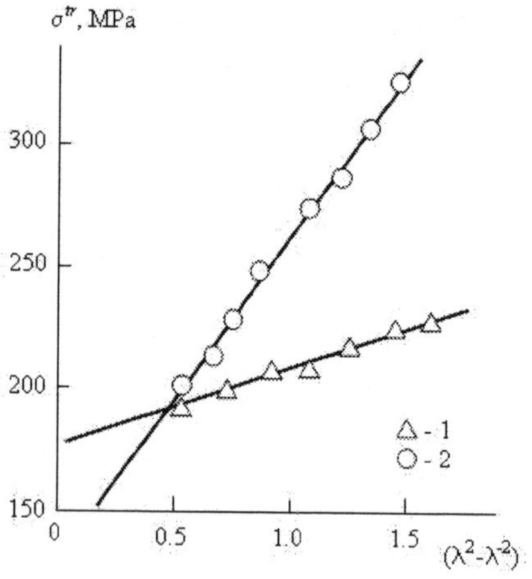

Figure 3.35. The dependences of true stress σ^{tr} on parameter $(\lambda^2-\lambda^{-2})$ for carbon plastics on the basis of phenylone at l_f=40 mm, t=30 s (1) and l_f=40 mm, t=240 s (2) [61].

G_p value calculation by the indicated method shows that it varies within the limits of 24.9-134 MPa. This is substantially higher than the value for amorphous glassy and semicrystalline polymers number, where the similar variation makes 1.3-35.3 MPa for 23 polymers. Such distinction can be explained within the frameworks of the cold flow model, proposed in paper [62] and modified by the authors [63]. As it was noted above, the cluster model of polymers amorphous state structure supposes [44, 49] an availability in the mentioned state of polymers of local order domains (cluster), consisting of several collinear densely-packed segments of different macromolecules (amorphous analogue of crystallite with stretched chains. These clusters are simultaneously multifunctional nodes of entanglements physical network, which are connected by transferred load ("tie") chains and are immersed into loosely-packed matrix. The model [44, 49] also supposes, that in a yield point loosely-packed matrix mechanical devitrification occurs, owing to which polymer on cold flow plateau behaves like rubber (see the equation (3.29)). Therefore, the polymer deformation mechanism on the diagram σ-ε indicated section represents it the motion of clusters, connected by "tie" chains, in devitrificated loosely-packed matrix [63]. It is obvious, that for the studied carbon plastics it is possible to suppose, that in cold flow process filler fibers will be included and it is expected, that they will offer larger resistance to motion, than clusters, since their sizes are essentially larger than the sizes of the latter and this should be resulted to G_p substantial growth. Within the frameworks of such treatment it can be supposed that G_p value two structural factors will be defined. At first, one of these factors will be interfacial regions relative fraction φ_{if}, which are connected with both filler fibers surface and polymeric matrix. It is obvious that φ_{if} increase should result to G_p growth. The second factor is the fibers orientation, obtained in composite processing in rotating electromagnetic field. It's obvious that, the better fibers orientation in respect to compression axis, the smaller G_p value is. The fibers orientation degree in polymeric matrix can be described quantitatively with the aid of fibers orientation factor η. In Figure 3.36 the dependence G_p on ratio (φ_{if}/η)

value is adduced, which corresponds to the stated above considerations. As one can see, the linear correlation is obtained, confirming the made above suppositions. One should pay attention to the fact, that the plot $G_p(\varphi_{if}/\eta)$ slope for carbon plastics, prepared with ferromagnetic particles by length 70 mm, higher is essentially than in cases of l_f=20 and 40 mm. For coincidence of linear dependences at all l_f values it is required that in the case l_f=70 mm either the value φ_{if} increases, or the value η reduces. It is obvious, that the interfacial regions are formed in carbon plastics samples processing and in virtue of this their increase in deformation process is improbable. However, in the mentioned process the violation of fibers orientation, connected through interfacial layer with polymeric matrix, is possible. For carbon plastics, prepared with ferromagnetic particles by length of 70 mm using, the average value φ_{if} is approximately twice smaller than in the case of l_f=20 and 40 mm. This assumes fiber-polymeric matrix bonds weakening and possible violation of fibers orientation, i.e. η reduction.

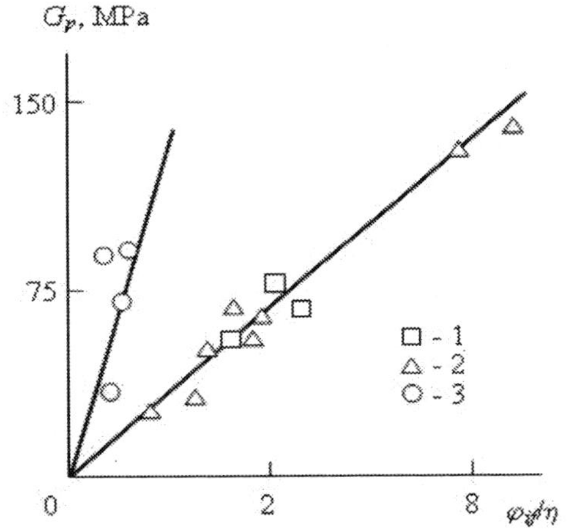

Figure 3.36. The dependence of strain hardening modulus G_p on ratio (φ_{if}/η) value for carbon plastics on the basis of phenylone at l_f=20 (1), 40 (2) and 70 mm (3) [61].

Let us consider in conclusion one more aspect of model [44, 49] application for behaviour description of carbon plastics on cold flow plateau. The densely-packed regions relative fraction can be calculated according to the equation (1.88). If in the indicated equation the value ν_{cl}, obtained according to the equations (3.29) and (1.87) is used, then φ_{dens} variation for the studied carbon plastics within the limits of 0.4-2.0 will be obtained. It is obvious, that for polymers (polymeric matrix) structure value φ_{dens}>1.0 has no physical significance. This means that according to the equation (1.88) the effective value φ_{dens} is obtained, considering the influence on parameter G_p of filler fibers motion on cold flow plateau.

Therefore, the results stated in the present section have shown that filler fibers motion in polymeric matrix in cold flow process is the strongest factor influencing on strain hardening modulus value. The indicated factor action can be expressed through ratio of relative fraction of the connected with fiber interfacial regions and fibers orientation factor. Fibers motion influence results to substantial growth of strain hardening modulus in comparison with nonfilled polymers [61].

3.4. The deformation Mechanisms and Plasticity of Polymer Composites

The ability to bear large strain with subsequent complete return at stress relieving is the property displayed at corresponding conditions by actually all polymeric substances, consisting of long chain macromolecules [64]. What is more, it is displayed exclusively by materials of such structure. This property is important one beyond narrow boundaries of "rubber elasticity" term, by which it is usually designated. The indicated property acts at polymeric networks swelling and at the deformation of substances, which are not included in elastomers category in general, for example, semicrystalline polymers, at viscoelastic behaviour of linear polymers in case of flow in liquid or amorphous states. The main prerequisite of rubber elasticity molecular theory serves the supposition that stress in rubbers is consequence of covalent network chains deformation, whereas the contribution of interaction between chains is negligibly small. Strictly speaking, this is not entirely correct even for true rubbers [64] and the more so for polymers in glassy state. The rubber elasticity theory gives the following expression for elastomers limiting draw ratio λ_f estimation [65]:

$$\lambda_f = n_{st}^{1/2}, \qquad (3.30)$$

where nst is a statistical segments number on chain section between chemical cross-linking nodes or physical entanglements.

To apply the considered conception to a glassy polymer number of empirical assumptions are made taking into account essentially stronger intermolecular interaction in such systems. Edwards and Vilgis [66] offered the slipping links conception, which supposes chain division between entanglements on smaller fragments, which are fixed, but have essential internal freedom. This results to polymers limiting strain reduction in comparison with the estimated one according to the equation (3.30). The authors [67] offered more general mode of the considered above accounting effects, based on the fractal analysis application for the description of deformation behaviour of carbon plastics on the basis of phenylone in compression tests.

In paper [68] the fractal relationship for limiting draw ratio λ_f calculation of polymeric materials was obtained:

$$\lambda_f = C_\infty^{D_{ch}-1}, \qquad (3.31)$$

where C_∞ is characteristic ratio, D_{ch} is fractal dimension of chain part between its fixation points (chemical cross-linking nodes, physical entanglements and so on). As it was noted above, the dimension D_{ch} changes within the limits 1-2 and characterizes molecular mobility level in polymers [69].

For comparison with theoretical estimations according to the equation (3.31) 11 curves stress-strain (σ-ε) for the studied carbon plastics were selected, for which clearly expressed failure point was observed, that allows the precise determination of experimental values λ_f. Two examples of such curves σ-ε are given in Figure 3.37. In Figure 3.38 the comparison of

λ_f and calculated by the indicated mode values λ_f^T is adduced. As one can see, a good correspondence of theory and experiment is obtained (the average discrepancy makes up ~ 5 %). This means, that carbon plastics failure in compression tests is controlled by chaining stretching between clusters and described within the frameworks of fractal mode [68].

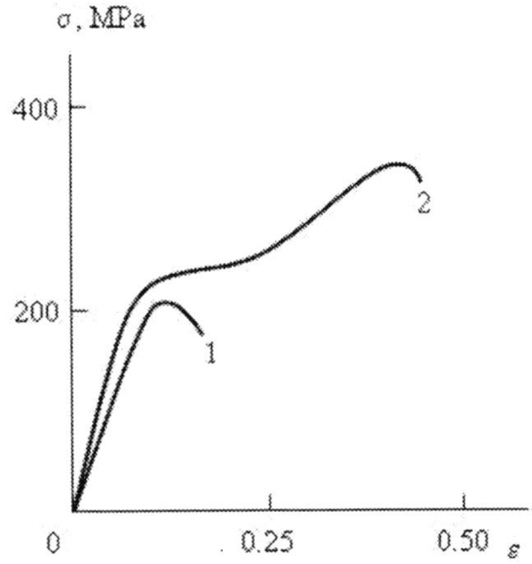

Figure 3.37. The curves stress-strain (σ-ε) for carbon plastics on the basis of phenylone at l_f=20 mm, t=20 s (1) and l_f=40 mm, t=180 s (2) [67].

Let us consider further deformation mechanisms and plasticity of composites with semicrystalline matrix. As a rule, filler introduction in semicrystalline polymers results to their plasticity (deformability) essential reduction [70]. Since polymer composites are structurally complex solids in general and this factor is increased by the usage of semicrystalline polymeric matrix, structure of which is complex enough by itself, then quite enough plasticity conceptions for these materials exist, considering the mentioned property from different positions [17, 70]. At deformation description of such composites it is necessary to consider crystalline morphology change at filler introduction, interfacial effects on border polymer-filler, short fibers aggregation and orientation in polymeric matrix and so on. However, one more factor exists, to which no attention was practically paid before – polymeric matrix structure change on molecular and supersegmental levels at filler introduction. This factor influence study within the frameworks of the cluster model of polymers amorphous state structure [44] and fractal analysis [5] shows that polymeric matrix structure modification in comparison with initial matrix polymer plays the main role in polymer composites properties change and the filler role comes to this modification realization and its fixation. Proceeding from this, the authors [71] studied the dependence of carbon plastics on the basis of HDPE plasticity on polymeric matrix structural characteristics of nanometer scale.

Carbon fibers contents increase in carbon plastics on the basis of HDPE results to their plasticity sharp reduction and deformation mechanism change at φ_f>0.08, as follows from the study of the fractured samples photographs, adduced in Figure 3.39. At small φ_f deformation development in carbon plastics is similar to this process in HDPE – the samples are deformed up to large draw ratio λ with necking that is usual for such polymers [72]. At φ_f>0.08 the failure has typically brittle character with λ≤1.3 (Figure 3.39).

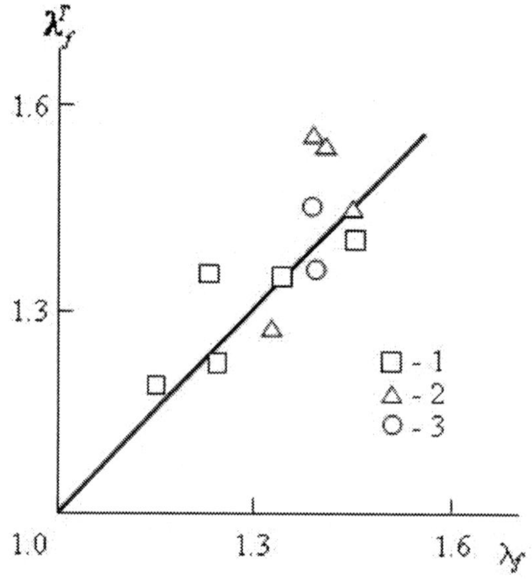

Figure 3.38. The comparison of experimental λ_f and calculated according to the equation (3.31) λ^T_f limiting draw ratio values for carbon plastics on the basis of phenylone at l_f=20 (1), 40 (2) and 70 mm (3) [67].

Within the frameworks of fractal analysis it was shown [7] that solids fracture type is controlled by their structure characterized by fractal (Hausdorff) dimension d_f. At d_f=2.50 the failure character is brittle, at 2.50<d_f<2.67 – quasibrittle (quasitough) and at d_f>2.70 – tough. The calculation of the dimension d_f according to the equation (1.1) was fulfilled for application correctness estimation of this criterion to the studied carbon plastics failure. In Figure 3.40 the dependence of limiting draw ratio λ_f on d_f value for carbon plastics on the basis of HDPE is adduced. As one can see, at d_f≤2.67 brittle (λ_f≤1.30) fracture is observed and at d_f>2.70 fracture type changes on the tough one, that accompanies by λ_f sharp growth up to λ_f greatest value ~ 6. Therefore, proposed in [7] structural criterion of solids plasticity is completely applicable in case of carbon plastics on the basis of HDPE.

Let us consider carbon plastics deformation mechanisms and their dependence on structural factors of nanometer scale. The authors [73] supposed that drawing occured as a result of crystalline and amorphous molecular sequences straightening and in this case the value λ_f can be expressed in times number f, which macromolecule passes through the same crystallite or cluster:

$$\frac{1}{\lambda_f}=\frac{K}{f}+\frac{(1-K)^{1/2}}{n_{st}^{1/2}}, \qquad (3.32)$$

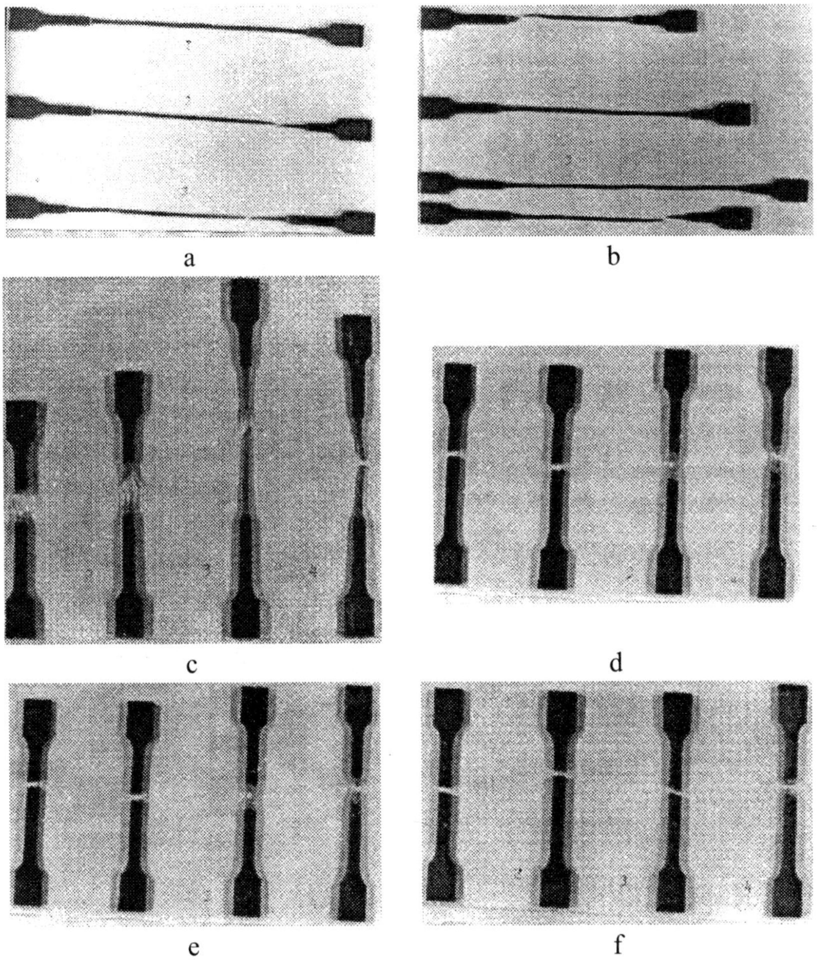

Figure 3.39. The photographs of carbon plastics on the basis of HDPE with filling volume degree φ_f: 0.038 (a), 0.076 (b), 0.114 (c), 0.152 (d), 0.227 (e), 0.303 (f), deformed at tests temperatures 293 (1), 313 (2) and 353 K (3) [71].

where K is crystallinity degree, n_{st} is an equivalent statistical links number between molecular entanglements nodes in melt. The value n_{st} varies usually from 100 up to 300; since this value does not substantially influence on results, then it was accepted equal to 225 [74]. The authors [74] were shown that in case of semicrystalline HDPE the value f was large and approximately equal to 50 and in case of amorphous glassy polymers f<1, that defined limiting values λ_f distinction at these materials failure. In the last case clusters, which are amorphous analogue of crystallites with stretching chains, play role of deformable structural units [44]. Therefore, the stated results suppose that at f>1 the straightening of crystalline sequences from folding crystallites is the main deformation mechanism on nanometer scale and at f<1 – the drawing of chains sections between clusters. In Figure 3.41 the dependence of λ_f on f is adduced, from which sharp increase of this dependence slope follows at f≈1. Hence, λ_f reduction at φ_f>0.08 is defined by deformation mechanism change – at φ_f<0.08

crystalline phase is deformed and at φ_f>0.08 noncrystalline regions deformation becomes the prevalent one, that has strong restrictions (f<1) on λ_f value.

Polymeric materials deformability associates very often with molecular mobility [75]. As it was noted above, within the frameworks of fractal analysis molecular mobility level can be characterized with the aid of fractal dimension D_{ch} of chain part between it fixation points. The value D_{ch} is connected with physical and/or chemical bonds density according to the following equation [29]:

$$D_{ch} = \frac{\ln n_{st}}{\ln(4-d_f) - \ln(3-d_f)}, \tag{3.33}$$

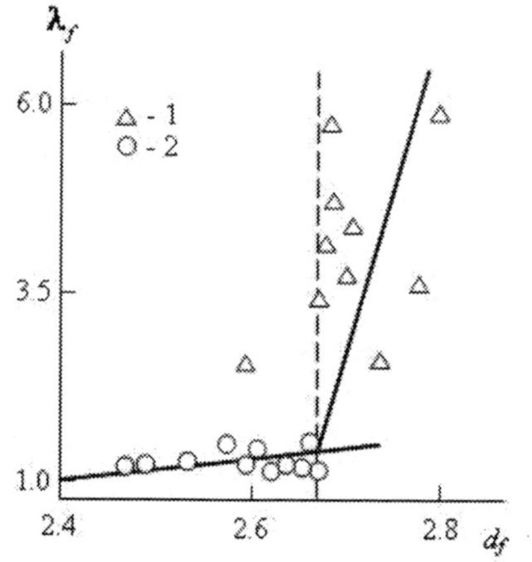

Figure 3.40. The dependence of limiting draw ratio λ_f on structure fractal dimension d_f for carbon plastics on the basis of HDPE in case of brittle (1) and tough (2) fracture types [71].

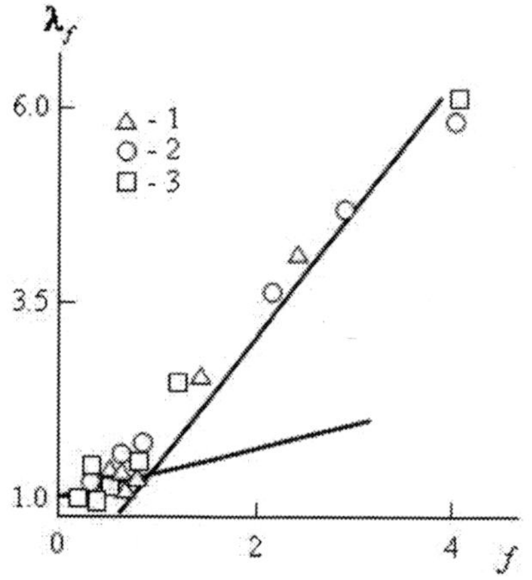

Figure 3.41. The dependence of limiting draw ratio λ_f on folding parameter f for carbon plastics on the basis of HDPE at testing temperatures 293 (1), 313 (2) and 353 K (3) [71].

where n_{st} is a chain statistical segments number between its fixation points.

In Figure 3.42 the dependence f on value n_{st}, calculated from combination of the equations (1.16), (1.55) and (3.33), is adduced (it is obvious, that n_{st} increase means macromolecular entanglements network density reduction). As it follows from the data of Figure 3.42, n_{st} the increase results to f growth and at $n_{st}\approx 9$ the dependence $f(n_{st})$ slope sharp increase begins, that should result to carbon plastics deformability strong enhancement (Figure 3.41). Let us note, that in case of cross-linked polymers the value $n_{st}=9$ divides densely cross-linked ($n_{st}<9$) and weakly cross-linked ($n_{st}>9$) networks [39]. The data of Figure 3.42 assume that this classification is true for physical entanglements network as well. Hence, for straightening of crystalline sequences from folded crystallites it is necessary to fulfill supplementary condition – physical entanglements network density should be small enough in order not to hinder this mechanism realization at carbon plastics deformation.

Hence, the stated above results demonstrated plasticity (deformability) strong dependence of carbon plastics on the basis of HDPE on polymeric matrix structural characteristics of a nanometer scale. In general terms these materials plasticity is controlled by their structure fractal dimension. Besides, high values of limiting draw ratio can be realized only at crystalline phase deformation and macromolecular physical entanglements network small density, allowing this deformation mechanism realization.

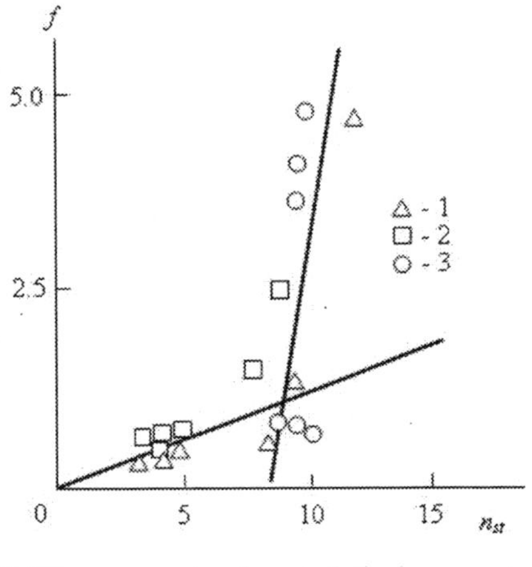

Figure 3.42. The dependence of folding parameter f on statistical segments number on chain section between clusters n_{st} for carbon plastics on the basis of HDPE at testing temperatures 293 (1), 313 (2) and 353 K (3) [71].

The authors [68] modified the equation (3.30) by the introduction of exponent (D_{ch}-1) in it instead of constant value ½:

$$\lambda_f = n_{st}^{D_{ch}-1}. \tag{3.34}$$

In papers [76, 77] this equation was used for the description of carbon plastics on the basis of HDPE deformability. In Figure 3.43 the dependence $\lambda_f(n_{st})$ is adduced, from which λ_f increasing at n_{st} growth follows, that was expected according to the equations (3.30) and (3.34). It is interesting to note, that this dependence breaks into two parts with sharply distinguishing slopes, moreover this transition realizes at $n_{st}\approx 9$. The reasons of indicated transition have been considered above. Let us point out, that the shape of the dependence

$\lambda_f(n_{st})$, adduced in Figure 3.43, exludes the application possibility of the equation (3.30) with constant exponent ½ for its correct description [76].

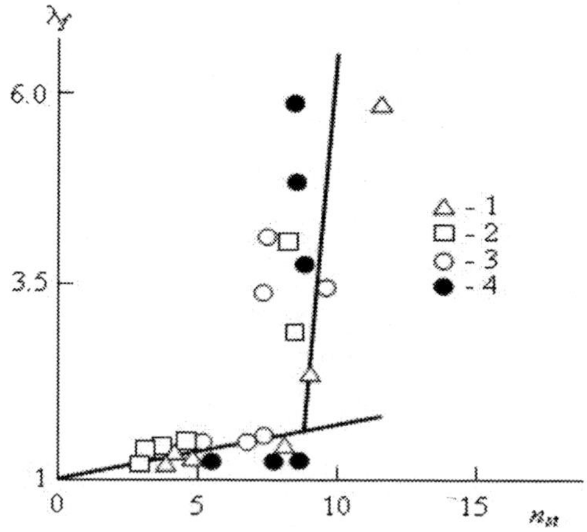

Figure 3.43. The dependence of limiting draw ratio λ_f on statistical segments number n_{st} on chain section between clusters for carbon plastics on the basis of HDPE at testing temperatures 293 (1), 313 (2), 333 (3) and 353 K (4) [77].

In Figure 3.44 the comparison of experimental λ_f and calculated according to the equation (3.34) λ_f^T limiting draw ratio values for the studied carbon plastics at four testing temperatures is adduced. As one can see, between λ_f and λ_f^T a good correspondence is obtained and the observed data scatter in respect to the straight line, giving relation 1:1, is symmetrical one and is due to well-known failure process statistical character [78].

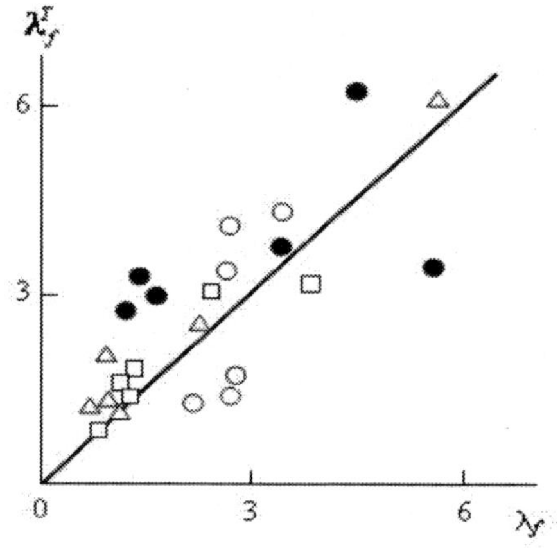

Figure 3.44. The comparison of experimental λ_f and calculated according to the equation (3.34) λ_f^T limiting draw ratio values for carbon plastics on the basis of HDPE. The straight line gives relation 1:1. The designations are the same as in Figure 3.43 [77].

As it is known [79], the fracture work U_f, characterizing energy expenditures on material deformation up to fracture, is one of the most important plasticity characteristics. It has been shown above that solids fracture character is defined by their structure fractal dimension d_f: at d_f=2.50 the brittle fracture is realized at d_f=2.50-2.67 – quasibrittle (quasitough) and at d_f>2.70 – tough fracture [7]. Such classification allows to suppose the plasticity increase, characterizing by value U_f, at d_f growth. Actually, the dependence $U_f(d_f)$, adduced in Figure 3.45, confirms this supposition. The indicated dependence is linear, U_f increasing at d_f growth is observed and zero value U_f is reached at d_f=2.50, i.e. at brittle fracture. Since the limiting (the greatest) value d_f for real solids is equal to 2.95 [7], then this allows to estimate the greatest value U_f for the studied carbon plastics, which is equal to ~ 17 MJ. The data of Figure 3.45 scatter is due again to fracture process statistical character [76, 77].

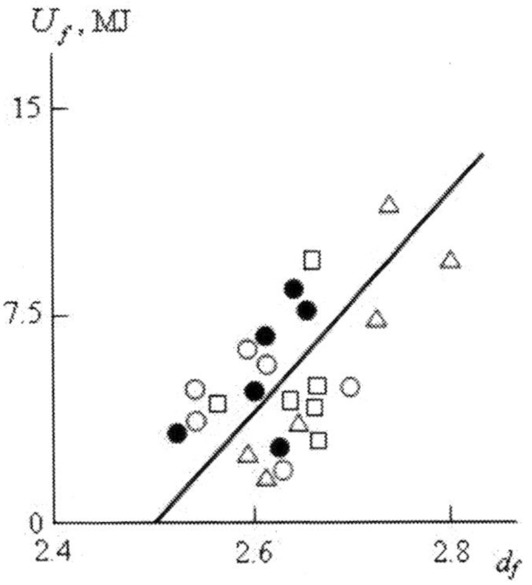

Figure 3.45. The dependence of fracture work U_f on structure fractal dimension d_f for carbon plastics on the basis of HDPE. The designations are the same as in Figure 3.43 [77].

Therefore, the stated above results showed that fractal geometry of chain section between its fixation points to a considerable extent defines carbon plastics deformability (plasticity). This is true with regard to density of physical entanglements cluster network, which also influences on the value D_{ch}. Carbon plastics plasticity is controlled by polymeric matrix structure state, characterized by its fractal dimension.

3.5. FRACTURE PROCESS OF POLYMER COMPOSITES

As it is well-known [80], elastic solids can be destroyed at compression. Besides the fracture has often a pillared character and solid division on vertical pillars, formed by cracks, growing in uniaxial compression direction [80]. Such character of fracture at compression is observed for polymer composites filled with short fibers [24]. This effect is not in conformity with theoretical representations of fracture classical mechanics. From the point of view of traditional theory stress intensity factor at crack (one-dimensional cut), oriented along compression direction, is equal to zero. Since this factor is the main parameter, characterizing fracture, then according to the classical notions such crack can not be propagated, that is in contradiction with experimental results [81].

For this contradiction settlement the authors [81] offered a fractal model of fracture at compression, which is based on a well-known fact of crack surface fractal structure [82], including polymers fracture [83]. The authors [84] applied the fractal model [81] for the description of fracture of carbon plastics on the basis of phenylone in uniaxial compression tests.

The dependence of fracture stress σ_f^c at compression on components blending duration in rotating electromagnetic field t for carbon plastics on the basis of phenylone is shown in Figure 3.46. It is significant that its shape is completely similar to the shape of the dependence of carbon plastics structure fractal dimension on t (see Figure 1.1). As it was noted above, such shape of dependence σ_f^c or d_f on t is specific for synergetic structures: at first periodical (ordered) behaviour of d_f or σ_f^c, close to a sigmoid one with period doubling, is observed and then transition to chaotic behaviour is realized. The indicated analogy supposes correlation existence between structural factor d_f and carbon plastics property σ_f^c. Actually, the adduced in Figure 3.47 plot confirms this assumption: the dependence σ_f^c (d_f) turns out to be linear and showing σ_f^c growth at d_f increasing, that analytically is described by the following relationship [84]:

$$\sigma_f^c = 190 + 360(d_f - 2), \text{ MPa}. \tag{3.35}$$

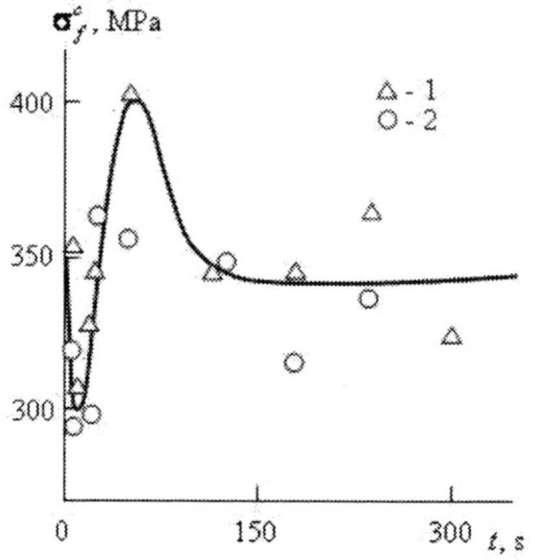

Figure 3.46. The dependence of fracture stress σ_f^c on components blending duration in rotating electromagnetic field t for carbon plastics on the basis of phenylone, prepared with magnetic (1) and mechanical (2) separation application [84].

The adduced in Figure 3.47 correlation has an empirical character and demonstrates the possibility of carbon plastics strength change within wide enough limits (293-406 MPa) at using of the described above mode of their preparation by means of variation of components blending duration in rotating electromagnetic field.

The fractal model [81] allows more precise description of carbon plastics fracture at compression. In papers [82, 85] on the example of crack propagation in tension conditions it was shown, that crack surface fractal structure account results to a change of stress σ_f^T asymptotic behaviour in its top neighbourhood. The relationship [81] is true for fractal cracks:

$$\sigma_f^T \sim K_I r^{\alpha}, \tag{3.36}$$

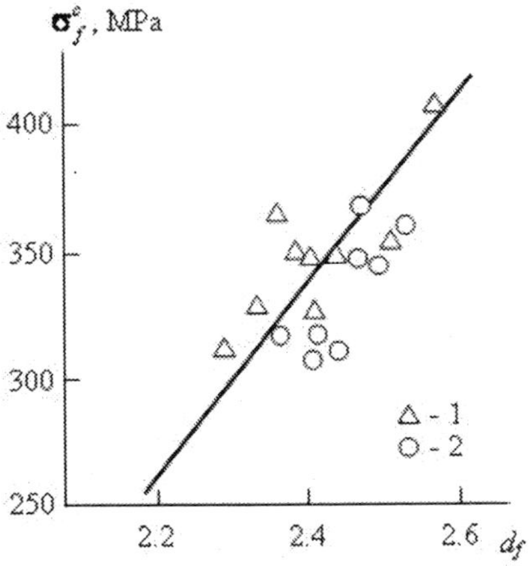

Figure 3.47. The dependence of fracture stress σ_f^c on structure fractal dimension d_f for carbon plastics on the basis of phenylone. The designations are the same as in Figure 3.46 [84].

where K_I is a dimensional proportionality coefficient in power asymptotics of stresses field, which is usually named as stress intensity factor, r is distance from crack top.

If as the crack model isotropic fractal is accepted, then it can be written [81]:

$$\alpha = \frac{d_{cr} - 2}{2}, \tag{3.37}$$

$$K_I \sim (d_{cr} - 1)^{1/2}, \tag{3.38}$$

where d_{cr} is fractal dimension of crack (fracture) surface ($2 \le d_{cr} \le 3$).

Since the considered carbon plastics are destroyed beyond yield stress, then formula for tough fracture type was used for value d_{cr} calculation [7]:

$$d_{cr} = \frac{2(1+4\nu)}{1+2\nu}, \tag{3.39}$$

where ν is Poisson's ratio.

Further, accepting the value r in the relationship (3.36) equal to 10 relative units, the value σ_f^T can be calculated. In Figure 3.48 the comparison of theoretical σ_f^T and experimental σ_f^c fracture stress values at compression for both series of the studied carbon plastics is adduced. As one can see, between σ_f^T and σ_f^c a good enough linear correlation is obtained, confirming fractal model [81] application correctness for carbon plastics on the basis of phenylone fracture description. This relationship is expressed analytically as follows [84]:

$$\sigma_f^c = 100\left(\sigma_f^T + 1\right), \text{ MPa.} \tag{3.40}$$

Let us note, that within the frameworks of fractal models fracture stress can be directly connected with polymeric materials structural characteristics (for example, d_f), but not with their secondary structures (fracture surfaces). With this purpose it is easily obtained from the equations (1.1) and (3.39) [84]:

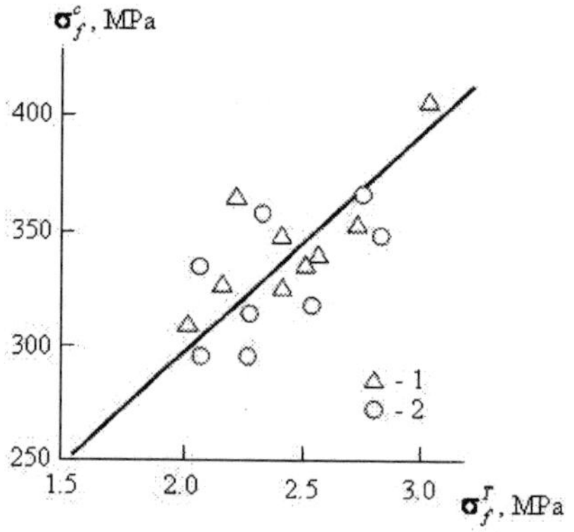

Figure 3.48. The comparison of experimental σ_f^c and calculated according to the fractal model σ_f^T fracture stress values for carbon plastics on the basis of phenylone. The designations are the same as in Figure 3.46 [84].

$$d_{cr} = \frac{2(2d_f - 3)}{d_f - 1}, \tag{3.41}$$

and the from combination of the equations (3.37), (3.38) and (3.41) can be written [84]:

$$\alpha = \frac{d_f - 2}{d_f - 1}, \tag{3.42}$$

$$K_I \sim \left(\frac{3d_f - 5}{d_f - 1}\right)^{1/2}. \tag{3.43}$$

From the principal point of view the equation (3.41) is important, since it describes the direct interconnection structure-property in fracture process. Let us consider several limiting cases. So, for rubbers (d_f=3) in the tough fracture case let us obtain d_{cr}=3, i.e. fracture surface will be very rough (porous). For polymer with the greatest ordering degree (d_f=2) d_{cr}=2, i.e. fracture surface will be smooth. Such interconnection can be obtained for brittle fracture, for which d_{cr} value is determined according to the equation [7]:

$$d_{cr} = \frac{10(1+\nu)}{7-3\nu}. \tag{3.44}$$

In this case let us obtain [84]:

$$d_{cr} = \frac{10d_f}{20-3d_f}. \tag{3.45}$$

Then at d_f=2.5 (the ability to ideal brittle fracture materials with $\nu \le 0.25$ have [7]) let us obtain d_{cr}=2, and at d_f=3 (rubbers) $d_{cr} \approx 2.73$.

Therefore, the assumption about fracture fractal character of the studied carbon plastics allows to eliminate the noted above experiment and classical theory, in which cracks are simulated by smooth (d_{cr}=2) mathematical cuts, discrepancy. In addition fractal model gives a good correspondence with experiment. Let us note that in paper [84] the indicated model is used successfully for composites tough fracture description.

As it was noted above, the strength of polymer composites filled with short fibers is defined to a considerable extent by these fibers length. The formula, derived for composites strength calculation in assumption of stresses homogeneous field along fibers, was modified for the account of two limiting cases [17]. In the first of these cases fibers have average length $\bar{l}$ more than critical length l_c and can be loaded up to fracture. In this case composite strength σ_f^c is calculated according to the formula (3.12). In the second case the fiber average length is smaller than the critical one ($\bar{l}<l_c$) and σ_f^c calculation is fulfilled according to the equation (3.13). The authors [86] fulfilled the study of fiber length and its change in preparation process influence on strength of composites on the basis of polyarylate (PAr).

In Figure 3.49 the dependences of strength of composites on the basis of PAr, filled with three types of fibers (uglen, vniivlon and glassy fiber), σ_f^c on their volume contents φ_f are shown. As one can see, these dependences distinction has not only quantitative character, but also qualitative one – if for PAr-vniivlon and PAr-glassy fiber the value σ_f^c grows at φ_f increasing then for PAr-uglen the opposite picture is observed. To elucidate the cause of such distinction let us consider the plot of Figure 3.10, where the dependence of strength σ_f^c on fiber average length $\bar{l}$ for composites on the basis of PAr is shown. In this case the dependence is the same for all three studied composites – at small $\bar{l}$ sharp enough enhancement σ_f^c is observed and then this dependence reaches plateau and certain reduction σ_f^c at $\bar{l}$ growth is even observed [19].

For fiber critical length l_c estimation the authors [19] were used the equation (3.12) and experimental data for composite PAr-glassy fiber with the greatest value $\bar{l}$=360 mcm. Then l_c is equal to 252 mcm. This l_c value is indicated on Figure 3.10 by vertical shaded line. As one can see, this line actually separates the dependence $\sigma_f^c(\bar{l})$ on the two mentioned above sections, that confirms its calculation correctness. In Figure 3.10 the theoretical dependence $\sigma_f^c(\bar{l})$, calculated according to the equation (3.12) at l_c=252 mcm and σ_f^m=168 MPa (continuous line) is also adduced, which describes well the experimental data (points). Hence, these results suppose that fiber average length $\bar{l}$ is one of the main factors, defining the considered composites strength.

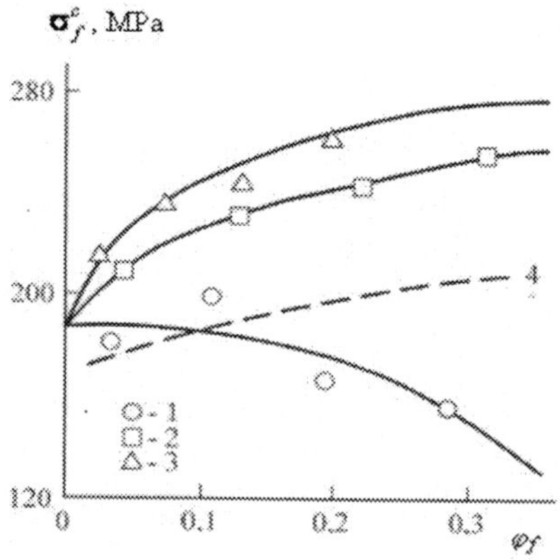

Figure 3.49. The dependence of fracture stress σ_f^c on filler volume contents φ_f for PAr-uglen (1, 4), PAr-vniivlon (2) and PAr-glassy fiber (3). 1-3 – the experimental data; 4 – calculation according to the equation (3.12) [86].

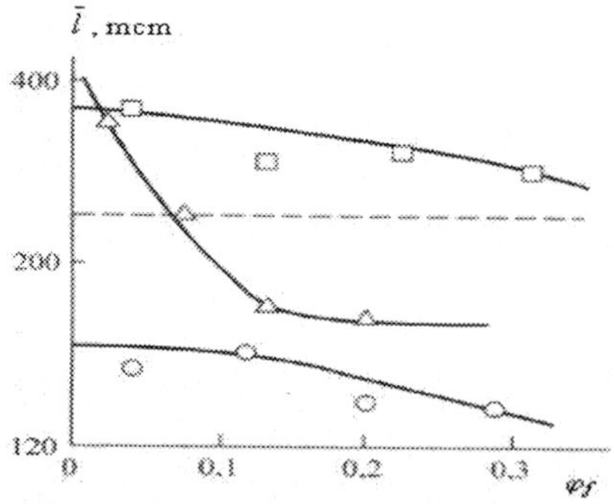

Figure 3.50. The dependence of fiber average length $\bar{l}$ on filler volume contents φ_f for composites in the composites on the basis of PAr. The horizontal shaded line points out fiber critical length l_c. The designations are the same as in Figure 3.49 [86].

For composites PAr-uglen the smaller values σ_f^c are due to the initial choice $\bar{l}$ for the initial fiber (Figure 3.50), where $\bar{l}$=170 mcm. For glassy fiber and organic fiber vniivlon the initial conditions are different. For them initial value $\bar{l}$ is larger than l_c, but if for vniivlon it is not reduced lower than l_c in the used φ_f interval, then for more brittle glassy fiber the condition $\bar{l}<l_c$ is already reached at small φ_f of order 0.08. In Figure 3.49 theoretical dependence σ_f^c (φ_f) for PAr-uglen is shown by the shaded line, which was calculated according to the equation (3.12) at the condition uglen fiber length equality to the corresponding parameter for vniivlon (both initial and changing in preparation process). One can see, that in this case σ_f^c values for PAr-uglen are no larger than for the obtained experimentally ones, but the dependence σ_f^c (φ_f) course changes, i.e. σ_f^c increase at φ_f growth is observed. Hence, for composites PAr-uglen the relatively small values σ_f^c are due to uglen fiber low strength (~ 500 MPa). For vniivlon fiber length optimal choice was made, submitting at all conditions to criterion $\bar{l}>l_c$ and relatively high strength of these fibers (850-1300 MPa) was obtained, that jointly results to stably high values of strength for composite PAr-vniivlon. And at last, the glassy fibers high strength defines the same stably high values of strength for PAr-glassy fiber at φ_f>0.08. Let us note, that vniivlon is the most suitable for preparation of composites with φ_f>0.3 and glassy fibers, subjected to strong reducing to fragments in preparation process, application will give σ_f^c reduction.

Hence, the stated above results showed that strength of composites on the basis of polyarylate, filled by short fibers of three types, was defined by fiber initial length, its reducing to fragments degree in preparation process and fiber strength. The organic fiber vniivlon on the basis of rigid-chain polyheroarylene possesses optimal properties of these criteria totality, that allows to recommend it for preparation of composites with filling high degree [86].

Mechanical properties of polymer composites, possessing good adhesion on interfacial boundary polymer-filler, are described according to the models, accounting the applied stress transfer possibility through this boundary by shearing mechanism. In the indicated case fracture stress σ_f^c of filled polymer will be a function of polymeric matrix strength σ_f^m and interfacial bond polymer-filler strength σ_a. These parameters interconnection is given by Leidner-Woodhams equation (the equation (1.42)) [87]. Though the indicated equation was obtained for filler spherical particles, it was derived on the basis of structural models for composites with short fibers and, as the authors [87] themselves noted, could be applied for such composites. The main lack of this equation is the difficulty of independent experimental determination of stress concentration coefficient K_s. Therefore the authors [88] undertook the attempt to physical significance definition of stress concentration coefficient in Leidner-Woodhams equation.

From the equation (1.42) it follows, that both σ_f^m and σ_a increase results to composites macroscopic strength σ_f^c growth. However, as it follows from the data of Figure 1.15, polymeric matrix and interfacial layers strength change is antibate, i.e. σ_a value reduces approximately linearly at σ_f^m increasing. Such dependence is due to feedback availability in carbon plastics on the basis of phenylone structure, which is a synergetic system, for which the relationship (1.10) is true. Hence, the value σ_f^c should have optimal magnitude at certain σ_a and σ_f^m. As experimental data showed, the greatest value σ_f^c, equal to ~ 406 MPa, was reached at high σ_a (~ 70 MPa) and relatively small σ_f^m (~ 120 MPa).

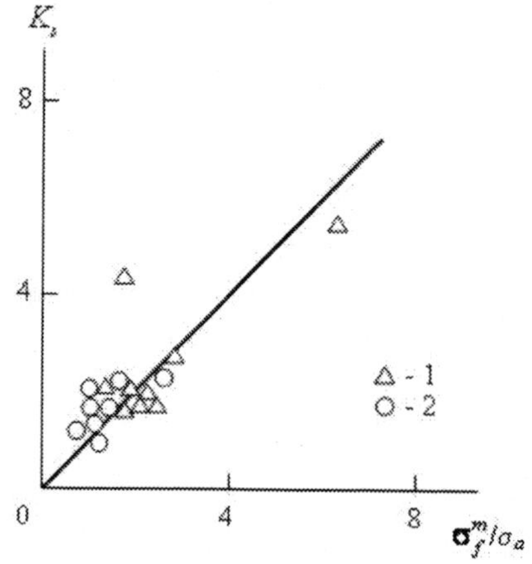

Figure 3.51. The relation between stress concentration coefficient K_s and ratio σ_f^m/σ_a for carbon plastics on the basis of phenylone, prepared with magnetic (1) and mechanical (2) separation application [88].

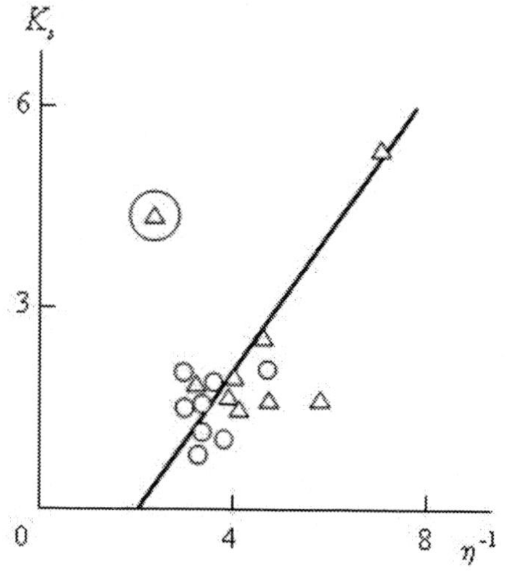

Figure 3.52. The dependence of stress concentration coefficient K_s on reciprocal value of fibers orientation factor η for carbon plastics on the basis of phenylone. The result for t=60 s is surrounded by small circle (the explanations are in the text). The designations are the same as in Figure 3.51 [88].

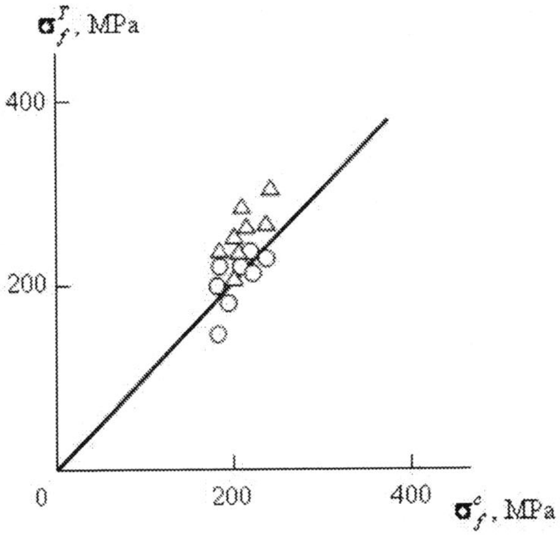

Figure 3.53. The comparison of experimental σ_f^c and calculated according to the equation (3.47) σ_f^T values of macroscopic strength of carbon plastics on the basis of phenylone. The designations are the same as in Figure 3.51 [88].

The value K_s calculation according to the equation (1.42) shows its considerable variation: K_s=0.96-5.49. The question arises about such variation cause in virtue of that circumstance, that for the studied carbon plastics φ_f=const, fibers geometry is not changed and their aggregation absence is supposed. It has been noticed that K_s value grows at σ_f^m increasing. Therefore it can be supposed that K_s is a function of the ratio σ_f^m/σ_a [87]. Actually, the adduced in Figure 3.51 dependence of K_s on σ_f^m/σ_a allows to approximate stress concentration coefficient as follows [88]:

$$K_s \approx \sigma_f^m / \sigma_a . \qquad (3.46)$$

From the equation (1.42) it follows that at constant σ_a, τ_m (or σ_f^m) and φ_f increase K_s results to composites macroscopic strength σ_f^c growth. Therefore the question arises, which structural parameter of carbon plastics controls the value K_s. As it is known [10], the fibers orientation factor η is a governing parameter for carbon plastics structure. In Figure 3.52 the dependence K_s on reciprocal value η is adduced, from which K_s decrease at η growth follows. However, the plotting of graph σ_f^c (η) has not discovered similar tendency, that assumes compensation of σ_f^m reduction by interfacial layer strength σ_a increasing (see Figure 1.15).

Combination of the equations (1.42) and (3.46) allows to obtain the following approximation of Leidner-Woodhams equation for carbon plastics [88]:

$$\sigma_f^c = \sigma_a + 1.48\sigma_f^m (1 - \varphi_f), \qquad (3.47)$$

where the stress σ_f^c is given for the case tension tests.

In Figure 3.53 the comparison of experimental σ_f^c and the calculated according to the equation (3.47) σ_f^T values of macroscopic strength for carbon plastics on the basis of phenylone is adduced. As one can see, between theory and experiment the good enough correspondence is obtained. This supposes, that the value K_s it is impossible the considered simply as a result of stress concentration by filler particles. For carbon plastics the value K_s, like in the work [87] is a ratio of bulk polymer matrix and interfacial layer strength. Let us note, that this rule violation, resulting to essential K_s growth (Figure 3.52) and σ_f^c, is obtained at components blending duration t=60 s, where negative feedback is observed [9] and, as consequence, the relationship (1.10) violation.

Hence, the considered above results showed applicability of Leidner-Woodhams equation for the description of strength of polymer composites, filled with short fibers. The stress concentration coefficient, as it was supposed originally, is a ratio of bulk polymeric matrix and interfacial layers strengths. The value of this ratio is defined by structure governing parameter-filler fibers orientation factor.

3.6. Polymer Composites Behaviour at Impact Loading

The impact toughness A_p of polymer composites is one of the most important mechanical characteristics, often defining their using possibility as engineering materials [2]. As a rule, impact energy dissipation occurs in the most loosely-packed regions of polymers structure [89]. However, for multiphase (personally, semicrystalline) polymers this problem is more complex [90]. So, for the mentioned semicrystalline polymers impact energy dissipates not only in vitrificated amorphous phase, but also in disordered in deformation process part of crystalline phase [90]. It is expected, that still more complex identification of regions, dissipating impact energy, will be in such heterogeneous and structurally complex materials as polymer composites. Therefore the authors [91] undertook clarification of polymer composites main structural components role in impact energy dissipation process on the example of carbon plastics on the basis of phenylone.

The main structural components of the studied carbon plastics are bulk polymeric matrix and interfacial regions, which, as a rule, differed are structurally (see chapter 1). In its turn, within the frameworks of the cluster model of polymers amorphous state structure [44, 49], the bulk polymeric matrix is defined as consisting of local order regions (clusters) and loosely-packed matrix. The authors [91] supposed that filler fibers did not participate in impact energy dissipation equally as densely-packed clusters, in which forming them macromolecules segments did not possess molecular mobility. Hence, on the role of impact energy dissipater in the studied carbon plastics two structural components aspire – interfacial regions and loosely-packed matrix with relative fractions φ_{if} and $\varphi_{l.m.}$, accordingly.

Within the frameworks of fractal analysis dissipated in impact loading process energy fraction η_e can be estimated as follows [92]:

$$\eta_e = 1 - \Lambda_n^{-\alpha}, \qquad (3.48)$$

where Λ_n is polymer structure automodelity coefficient, α is a fractional part of fracture surface fractal dimension d_{cr}.

For polymers $\Lambda_n = C_\infty$ (the equation (1.14)) and the value d_{cr} can be estimated with the aid of several methods, one of which is the equation using [90]:

$$d_{cr} = 2 - \frac{\ln(6 - 2d_f)}{\ln C_\infty}. \qquad (3.49)$$

In Figure 3.54 the comparison of value η_e and structural characteristics $\varphi_{l.m.}$, φ_{if} and their sum ($\varphi_{l.m.}+\varphi_{if}$) as well is also adduced. As one can see from this comparison, relative fractions of each from the indicated structural components do not discover correlation with η_e, but the sum ($\varphi_{l.m.}+\varphi_{if}$) corresponds precisely enough to the condition ($\varphi_{l.m.}+\varphi_{if}$)=$\eta_e$. Therefore, the data of Figure 3.54 indicate, that the impact energy dissipation is realized in two less densely-packed structural components of carbon plastics: loosely-packed regions of bulk polymeric matrix and interfacial regions.

However, the general correlation of impact toughness A_p and the dissipated energy fraction η_e could not be obtained (Figure 3.55). As one can see, this dependence is broken up into three linear correlations, passing through coordinates origin. For explanation of Figure 3.55 data the authors [91] used the thermal cluster model, which was described in detail in sections 1.2 and 1.3. Returning to Figure 3.55 data, let us note that different straight lines are controlled by various structural components of carbon plastics, that was found out from the indices β_T and β_p, ν_p, t_p comparison. The greatest values A_p at the same values η_e were obtained in the case, when the component, defining composite behaviour, is a cluster network, the smallest ones – for the loosely-packed matrix case and the intermediate ones – for the interfacial regions case. The general correlation $A_p(\eta_e)$ absence means more complex mechanism of impact energy dissipation than its dissipation with the aid of molecular mobility, as in a rubbers case [89, 90]. For glassy polymers shearing, the intensity of which grows at macromolecular entanglements network density increasing, is the most effective mechanism of dissipation [93]. This parameter for each of the mentioned structural components can be estimated as follows. The cluster network density ν_{cl} is determined according to the equation (1.88). The entanglements network density in interfacial regions ν_{if} can be also estimated according to the equation (1.88), if φ_{cl} is replaced on φ_{if} and C_∞=9 is accepted. Since in loosely-packed matrix only traditional entanglements ("fling-like" [44] or binary-hooking [94]) exist, then for phenylone such network density $\nu_{l.m.}$ can be obtained from literary sources [94-96]. The calculated by the indicated mode entanglements network density values as equal to: ν_{cl}=2.37-8.86×10^{27} m^{-3}; ν_{if}=0.43-2.26×10^{27} m^{-3}; $\nu_{l.m.}$=0.42×10^{27} m^{-3}. As one should expect, the cluster network discovers the greatest density, loosely matrix – the smallest one.

In Figure 3.56 the dependence of linear correlations $A_p(\eta_e)$ slope Δ according to the data of Figure 3.55 on entanglements networks density of different structural components is adduced. As one can see, Δ growth at network density increase is observed, that the made above supposition confirms. Therefore, the availability in the considered carbon plastics impact energy dissipation two mechanisms can be supposed further: the first of them dissipates energy at the expence of molecular mobility (typical for rubbers mechanism) and the second – at the expence of shearing zones formation. The second of the mentioned mechanisms is a typical one for plastic glassy polymers, to which phenylone belongs. For this postulate confirmation in Figure 3.57 the dependence A_p on product $\eta_e \nu_c$, characterizing combined intensity of both indicated mechanisms, is adduced. As it follows from the data of this Figure, the linear correlation $A_p(\eta_e \nu_c)$, passing through coordinates origin and common for all the studied carbon plastics, is obtained now.

Hence, the adduced above results have shown that loosely-packed regions of bulk polymeric matrix and interfacial regions are an impact energy dissipater in carbon plastics on the basis of phenylone. Impact energy dissipation is realized at the expence of two mechanisms: molecular mobility that is typical for rubbers, and shearing mechanism, typical for plastic glassy polymers. The macromolecular entanglements network density is a defining factor for the last mechanism.

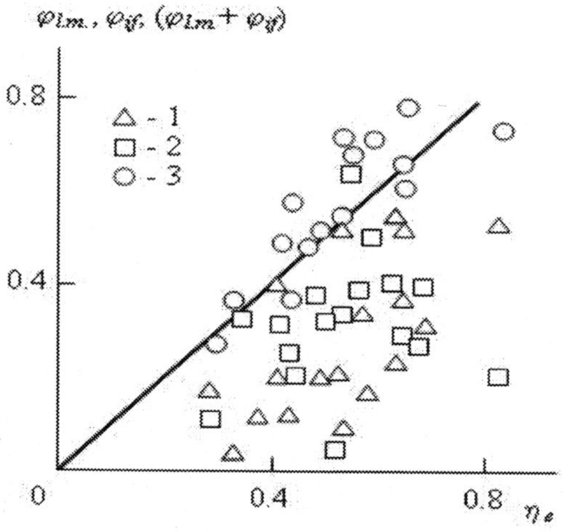

Figure 3.54. The comparison of relative fractions of loosely-packed matrix $\varphi_{l.m.}$ (1), interfacial regions φ_{if} (2), sum ($\varphi_{l.m.}$+φ_{if}) (3) and impact dissipated energy fraction η_e for carbon plastics on the basis of phenylone [91].

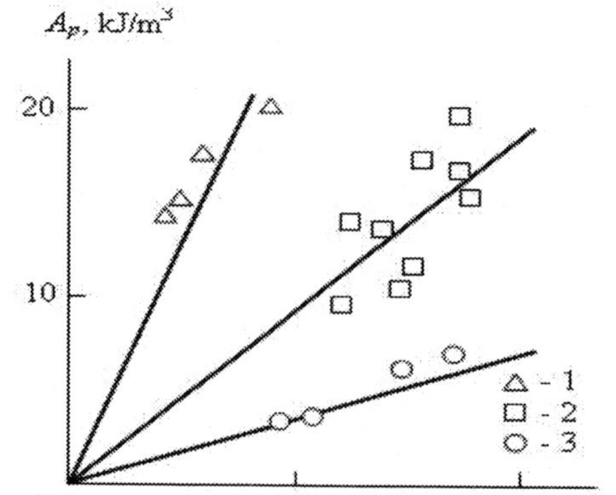

Figure 3.55. The dependences of impact toughness A_p on dissipated energy fraction η_e for carbon plastics on the basis of phenylone. The component, defining composite behaviour: cluster network (1), interfacial regions (2) and loosely-packed matrix (3) [91].

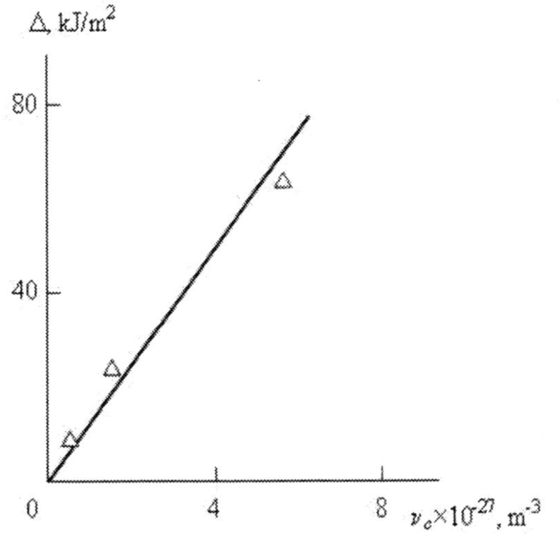

Figure 3.56. The dependence of plots $A_p(\eta_e)$ in Figure 3.55 slope Δ on macromolecular entanglements network density ν_c for carbon plastics on the basis of phenylone [91].

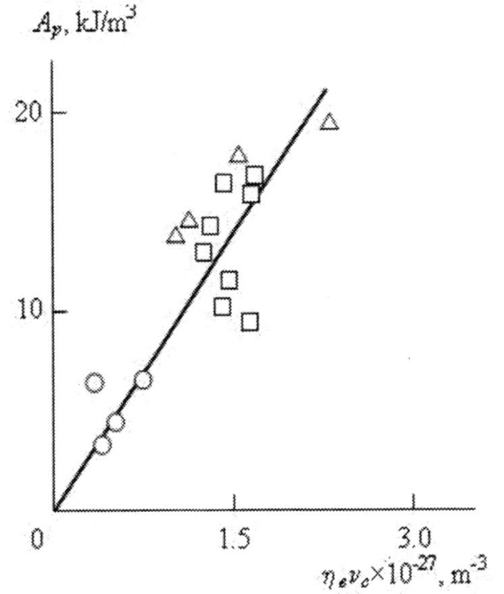

Figure 3.57. The dependence of impact toughness A_p on product $\eta_e\nu_c$ for carbon plastics on the basis of phenylone. The designations are the same as in Figure 3.55 [91].

The study of impact toughness for carbon plastics on the basis of phenylone within the frameworks of the thermal cluster model was continued in paper [97]. In Figure 3.58 the dependence of impact toughness A_p of carbon plastics on their components blending duration in rotating electromagnetic field t is shown. As it follows from the comparison of the dependences $\beta_T(t)$ and $A_p(t)$ (Figures 1.58 and 3.58, accordingly), the greatest A_p values correspond to periodic (ordered) composite structure behaviour and the smallest – to chaotic behaviour.

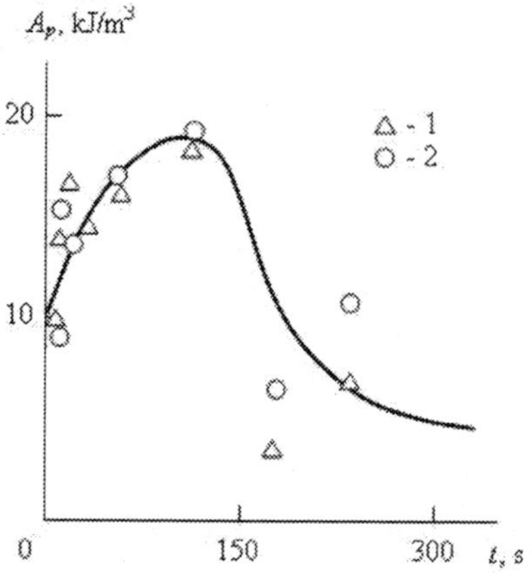

Figure 3.58. The dependence of impact toughness A_p on components blending duration in rotating electromagnetic field t for carbon plastics on the basis of phenylone, prepared with magnetic (1) and mechanical (2) separation application [97].

In Figure 3.59 the dependence $A_p(\beta_T)$ is shown, which has an extreme character. The smallest A_p values are reached at $\beta_T=\nu_p$, i.e. in that case, when loosely-packed matrix is a structural component, defining composite behaviour. The value A_p increases essentially at this defining role handing to cluster network or interfacial regions. Therefore, if densely-packed regions (clusters or interfacial regions) are structural component, defining composites behaviour, then the impact toughness values are relatively high and if loosely-packed matrix, consisting of chaotically tangled macromolecular coils, is this component, then A_p value is essentially lower. Let us note, that direct connection is observed between carbon fibers homogeneous (chaotic) distribution in polymeric matrix and structure chaos degree – at $t \geq 120$ s, when the indicated above distribution is reached, loosely-packed matrix, possessing chaotic structure, becomes the component, defining composite behaviour.

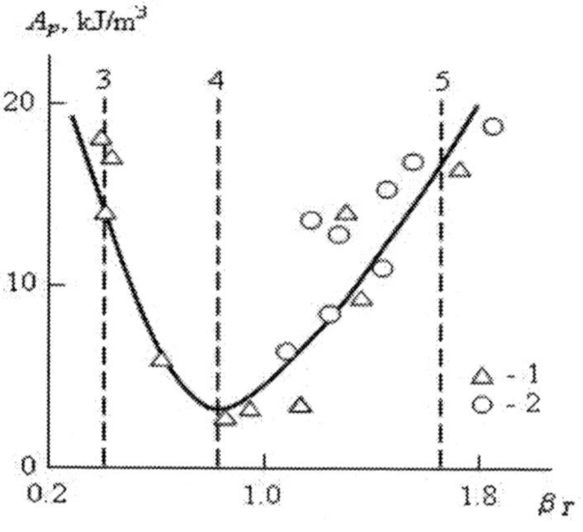

Figure 3.59. The dependence of impact toughness A_p on thermal cluster index β_T for carbon plastics on the basis of phenylone, prepared with magnetic (1) and mechanical (2) separation application. The vertical shaded lines point out percolation critical indices β_p (3), ν_p (4) and t_p (5) [97].

The similar picture is observed for particulate-filled composites PHE-Gr (Figure 3.60). As one can see, at $\beta_T > \nu_p = 0.74$ the essential A_p reduction occurs. As it is known [98], in these composites interfacial regions have low packing density in virtue of large fractal dimension value of filler particles aggregates surface. Therefore the extreme dependence $A_p(\beta_T)$ for composites PHE-Gr is not observed.

Therefore, the stated above results allow to suppose, that composites impact toughness value is defined by chaos degree in their structure.

One of the possible measures of chaos degree in system dynamic (in the given case – composite structure) is Lyapunov exponent λ_l [99]. It is a measure of exponential rate of neighbouring trajectories divergence or convergence in phase space along the given coordinate direction. Chaotic processes are characterized by neighbouring trajectories exponential divergence and in virtue of this – at any rate by one positive Lyapunov exponent. For λ_l estimation a general formula was used [100, 101]:

$$\lambda_l = \frac{\Delta S}{d_f}, \tag{3.50}$$

where ΔS is structural component entropy change, df is its fractal dimension.

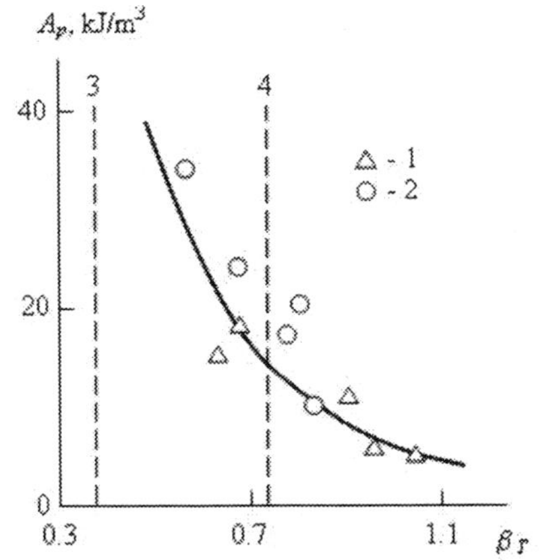

Figure 3.60. The dependence of impact toughness A_p on thermal cluster index β_T for composites PHE-Gr-I (1) and PHE-Gr-II (2). The vertical shaded lines point out percolation critical indices β_p (3) and ν_p (4) [97].

The values ΔS and d_f for the mentioned above structural components of carbon plastics were determined according to the equations (1.9) and (1.1), accordingly [97].

In Figure 3.61 the dependence of A_p on λ_l is adduced, which has an expected character: A_p value is reduced at λ_l growth. The greatest A_p value for the studied carbon plastics can be estimated by linear dependence $A_p(\lambda_l)$ extrapolation to $\lambda_l=0$, i.e. $\Delta S=0$ or $f_g=0$. This value is approximately equal to 25.5 kJ/m^2. Let us note one important observation. Plotting of the dependence A_p on total Lyapunov exponent, calculated for composite entire structure, shows the absence of correlation between these parameters. This confirms that A_p value of carbon plastics is controlled not by the entire structure, but by only one structural component.

At present is it well known, that either polymers property can be defined only by their structure part. So, yield process is controlled by structure densely-packed regions [48, 50], gas transport processes-loosely-packed regions [102] and so on. However, "change-over" effect of defining structural component for the same property was described in paper [97] for the first time.

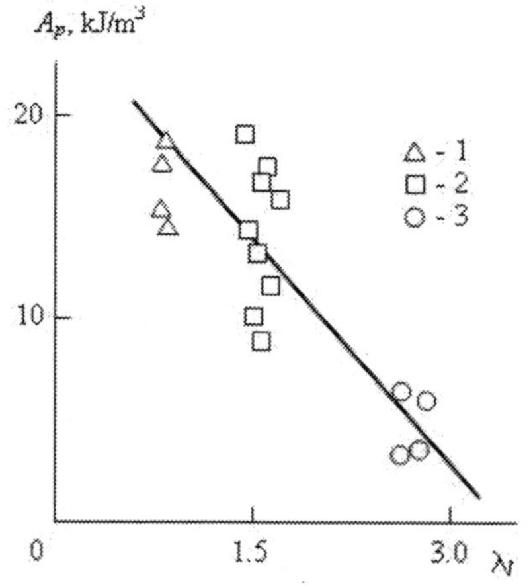

Figure 3.61. The dependence of impact toughness A_p on Lyapunov exponent λ_l for carbon plastics on the basis of phenylone. The controlling structural component: 1 – cluster network; 2 – interfacial regions; 3 – loosely-packed matrix [97].

Therefore, the stated above results have confirmed that the thermal cluster model allows to identify composites structural component responsible for impact toughness value of these materials. Densely-packed components promote A_p increase, loosely-packed ones – its reduction. This law can be quantitatively described by the dependence of A_p on structure chaos degree characteristic – Lyapunov exponent. From the practical point of view goal-directed "change-over" of controlling structural component allows to change essentially (in 5-7 times) A_p value [97].

As it is known [103], one of the synergetic systems main features is feedback availability and a specific characteristics number (governing parameter, feedback parameter and so on) for their description as well. Proceeding from this, the authors [104] studied the interconnection of impact toughness of carbon plastics on the basis of phenylone and their structure synergetic characteristics.

The feedback level in synergetic system can be characterized with the aid of feedback parameter λ, determined according to Poincaret's equation (the formula (1.12)). As it has been shown in paper [10], the carbon fibers orientation factor η is a governing parameter for carbon plastics structure (more precisely, for their interfacial regions). In Figure 3.62 the dependence A_p on feedback parameter λ for the considered carbon plastics is adduced. As one can see, the linear growth A_p at λ increase is observed, why at feedback absence (λ=0) A_p value is equal to zero at that, i.e. carbon plastics lose their plasticity completely. Therefore, polymeric material "pumping" from interfacial regions in densely-packed domains (clusters) of bulk polymeric matrix, characterized by λ value [9], is necessary condition of impact toughness high magnitudes reaching for the considered carbon plastics.

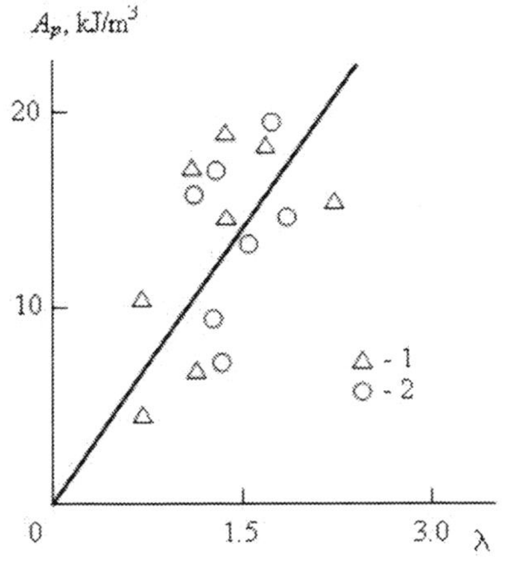

Figure 3.62. The dependence of impact toughness A_p on feedback parameter λ for carbon plastics on the basis of phenylone, prepared with magnetic (1) and mechanical (2) separation application [104].

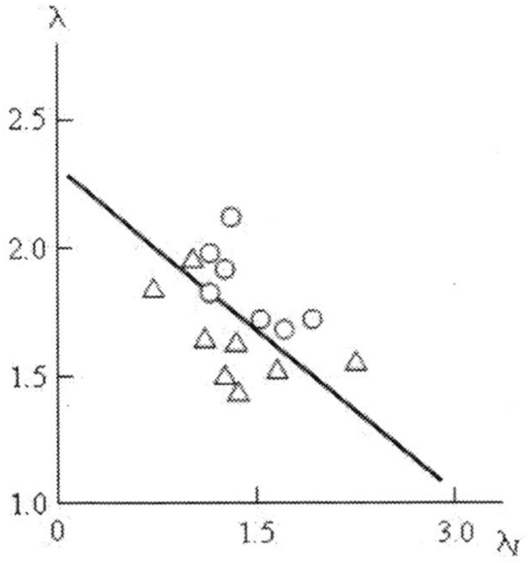

Figure 3.63. The relation between feedback parameter λ and Lyapunov exponent λ_l for carbon plastics on the basis of phenylone. The designations are the same as in Figure 3.62 [104].

As it was noted above [97], chaos level increase in carbon plastics structure, characterized by Lyapunov exponent λ_l, results to A_p reduction (see Figure 3.61). Therefore one can expect certain correlation between characteristics of feedback λ and chaos level λ_l.

In Figure 3.63 the dependence of λ_l on feedback parameter λ for both series of the studied carbon plastics is adduced. As it follows from the data of this Figure, feedback intensification, characterized by λ enhancement, results to chaos degree reduction, characterized by Lyapunov exponent, in carbon plastics structure.

In connection with the stated above results the question about regulation mode of feedback level or chaos degree in carbon plastics structure arises, i.e. the control mode of polymeric material "pumping" from one densely-packed structural component in to the other. It is obvious, that the natural candidate on such controlling role is the carbon plastics structure governing parameter – fibers orientation factor η [10]. Actually, the adduced in Figure 3.64 dependence $\lambda_l(\eta)$ turns out to be linear, showing λ_l growth at η increasing. Therefore, fibers orientation factor increase results to chaos degree growth and feedback level reduction in carbon plastics structure, that decreases A_p value (see Figure 3.58).

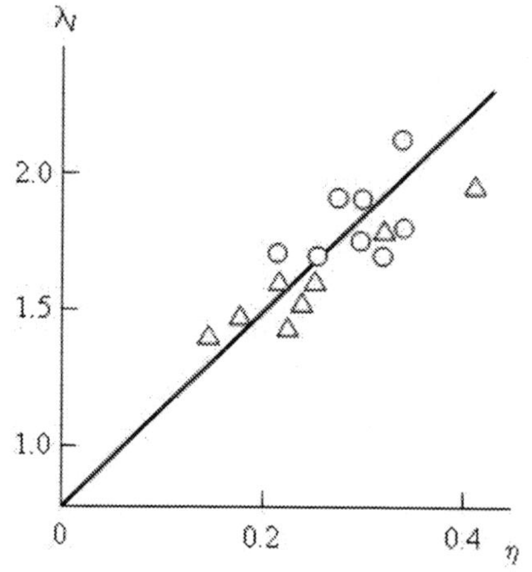

Figure 3.64. The dependence of Lyapunov exponent λ_l on fibers orientation factor η value for carbon plastics on the basis of phenylone. The designations are the same as in Figure 3.62 [104].

The dependence $\lambda_l(\eta)$ extrapolation to η=0 gives the value $\lambda_l \approx 0.86$. From the equations (1.1), (1.2), (1.8), (1.9) and (3.50) it follows that such value λ_l is reached at $d_f \approx 2.0$, i.e. in case of completely densely-packed structure of polymeric matrix, including interfacial regions too. The value η=0 in this case means the loss by fiber orientation factor of the governing parameter role. The latter is explained by the fact, that synergetics laws are applicable to nonequilibrium systems, whereas polymeric matrix structure at d_f=2.0 becomes thermodynamically equilibrium. However, in the real polymers case the condition d_f=2.0 is unattainable [33], therefore factor η will always be a governing parameter for real carbon plastics structure. Besides, from the equation (1.12) it follows, that at η_n=0 the value $\lambda \to \infty$ and from the plot of Figure 3.62 the condition $A_p \to \infty$, follows that for real polymers doesn't have physical significance.

Hence, the considered above data have demonstrated influence of the main characteristics of synergetic systems, to which carbon plastics on the basis of phenylone is applied, on their impact toughness. Increasing the feedback parameter, characterizing polymeric material "pumping" from interfacial regions in to bulk polymeric matrix, results to carbon plastics impact toughness growth and structure governing parameter (fiber orientation factor) enhancement – to its decreasing.

As it was noted above (see section 1.2), modern studies in surface chemistry and physics filed allow to describe processes, connected with surface energy role in structure and properties formation of nanometer size substance. The fractal dimension is a universal informant of substance structural state in both living and lifeless nature [105]. The topological dimension of volume d is equal to 3 and surface – to 2. The transition from volume to surface is characterized by surface layer fractal dimension d_f. In monograph [106] the conception of major properties of surface transitions from d=3 up to d=2, in conformity with which at reduction of three-dimensional space by substance filling dimension value at transition from material object bulk part on its surface energy is released, which is found experimentally as condensed phase surface energy. In addition surface energy value is defined by the difference d-d_f, where d is space topological dimension, equal to 3, and d_f changes within the limits $2 \le d_f < 3$. It is obvious, that any fractal structure gets under the given treatment, including polymer sample structure, which can be considered as transitional layer on boundary with the environment. Therefore the authors [71] attempted to establish interconnection between fractal structure of carbon plastics on the basis of phenylone and important mechanical parameter – effective surface energy, which characterizes material plasticity, within the frameworks of the conception [106].

For calculation of effective surface energy in the capacity of which critical deformation energy release rate G_{I_c} was used, the authors [71] applied the following methodics. As it is known [81], critical stress intensity coefficient K_{I_c} in compression tests is determined according to the equation (3.38) and fracture surface fractal dimension d_{cr} – according to the equation (3.44). Further the value G_{I_c} in relative units in virtue of proportionality, but not equality, sign in the relationship (3.38) can be calculated as follows [107]:

$$G_{I_c} = \frac{K_{I_c}}{E_c}. \qquad (3.51)$$

In Figure 3.65 the dependence G_{I_c}, calculated according to the equations (3.38), (3.44) and (3.51), on dimensions difference (d-d_f) is adduced, which has an expected character: d_f decreasing or (d-d_f) increasing results to G_{I_c} enhancement according to the mentioned above causes [105, 106]. Since in this case the values G_{I_c} are given in relative units, then for the plot G_{I_c} (d-d_f) for carbon plastics the following method can be used. In Figure 3.65 the dependence of the values G_{I_c} estimated experimentally [5] on (d-d_f) for particulate-filled composites PHE-Gr is adduced. As one can see, this plot is a linear one, and shows G_{I_c} enhancement at (d-d_f) increasing and at (d-d_f)=0 is extrapolated to the same value $G_{I_c}^0 \approx 0.24$ kJ/m^2, as the similar plot for carbon plastics. The plots for PHE-Gr and carbon plastics superposition allows to obtain the following analytic expression, giving interconnection of polymer composites structural characteristic d_f and plasticity G_{I_c} [71]:

$$G_{I_c} = 0.24 + 1.10\left(d - d_f\right), \text{ kJ/m}^2. \qquad (3.52)$$

Let us note, that for polymeric matrix Euclidean structure, i.e. at d_f=d or (d-d_f)=0 composites plasticity is finite and a minimum one ($G_{I_c}^0 \approx 0.24$ kJ/m^2) [71].

The value G_{I_c} estimations, fulfilled according to the equation (3.52), allow to calculate another important characteristic of composites structure – critical structural defect size a_{cr} – according to the equation [107]:

$$a_{cr} = \frac{G_{I_c} L}{72 A_p}, \tag{3.53}$$

where L is distance between impact device supports (span), Ap is impact toughness.

In Figure 3.66 the dependence of a_{cr} on components blending duration in rotating electromagnetic field t for carbon plastics on the basis of phenylone is shown. As one can see, within the range t=5-120 s the value a_{cr} is approximately constant and close to filler fibers diameter (7-9 mcm). This observation points out that critical structural defect formation in carbon plastics is the most probable on interfacial boundary polymer-filler, namely, over fiber butt-end surface. At t>120 s a_{cr} essential growth is observed supposing fibers aggregation possibility. Such course of the dependence $a_{cr}(t)$ the used for the given carbon plastics preparation the technology of their components preliminary blending in rotating electromagnetic field is defined. In Figure 1.1 the dependence of carbon plastics structure fractal dimension d_f on t is adduced, from which it follows that within the range t=5-120 s the approximately sigmoid dependence $d_f(t)$ is obtained, which at t≥120 s approaches to constant value d_f≈2.41. Such dependence type is specific for periodic (quasiperiodic) structures with the following system transition to chaotic behaviour (see section 1.1). This observation indicates, that the postulated above homogeneous (chaotic) carbon fibers distribution at the used method of components blending is not reached instantly, but only at t≥120 s, at the same time at t<120 s periodic (quasiperiodic) structures control carbon plastics behaviour [103]. Consequently, the comparison of Figures 1.1 and 3.66 data supposes that fibers aggregation can occur only at their chaotic distribution in polymeric matrix. This means, that minimum values a_{cr} and, hence, the greatest values A_p obtaining t range should be restricted by the limits of t=5-120 s (see Figure 3.58).

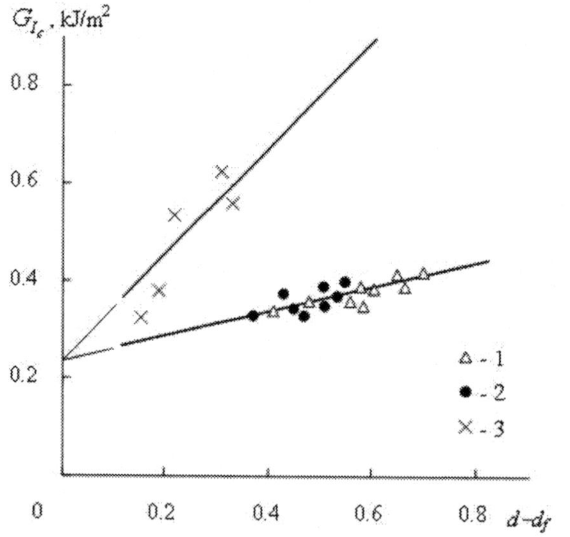

Figure 3.65. The dependence of critical deformation energy release rate G_{I_c} on difference (d-d_f) for carbon plastics on the basis of phenylone, prepared with magnetic (1) and mechanical (2) separation application and composites PHE-Gr (3) as well [71].

Hence, the stated above results have confirmed correctness of the conception, according to which the energy, discovered experimentally as surface energy, is released at the reduction of three-dimensional space filling by substance dimension. The obtained analytical expression allows to estimate quantitatively this energy only by structural characteristics. Fibers aggregation in carbon plastics is realized only at their chaotic distribution in polymeric matrix [71].

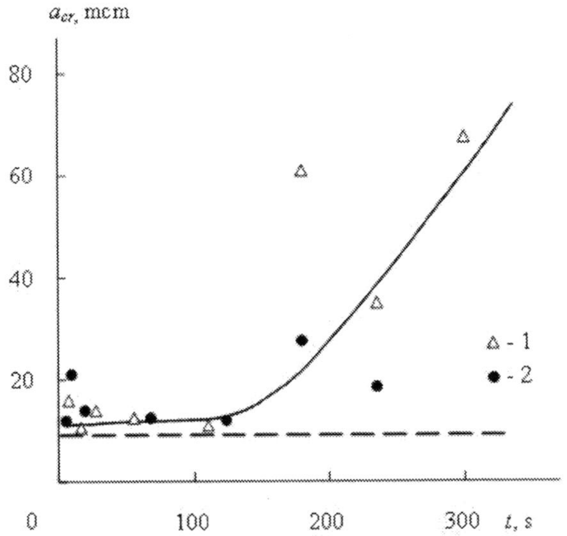

Figure 3.66. The dependence of critical structural defect size a_{cr} on components blending duration in rotating electromagnetic field t for carbon plastics on the basis of phenylone, prepared with magnetic (1) and mechanical (2) separation application. The value of carbon fiber diameter is indicated by a shaded line [71].

REFERENCES

[1] Bartenev G.M., Frenkel S.Ya. *The Physics of Polymers.* Leningrad, Khimiya, 1990, 432 p.

[2] Nilsen L.E. *Mechanical Properties of Polymers and Composites.* New York, Marcel Dekker, Inc., 1974, 309 p.

[3] Ahmed S., Jones F.R. *J. Mater. Sci.,* 1990, v. 25, № 12, p. 4933-4942.

[4] Bobryshev A.N., Kozomazov V.N., Babin L.O., Solomatov V.I. Synergetics of Composite Materials. Lipetsk, *NPO ORIUS,* 1994, 154 p.

[5] Novikov V.U., Kozlov G.V. *Mechanika Kompozitnykh Materialov,* 1999, v. 35, № 3, p. 269-290.

[6] Burya A.I., Kozlov G.V., Van'kov A.Yu. *Novosti Nauki Pridneprov'ya,* 2004, № 3, p. 33-36.

[7] Balankin A.S. Synergetics of Deformable Body. Moscow, *Publishers of Ministry Defence SSSR,* 1991, 404 p.

[8] Burya A.I., Kozlov G.V. *Voprosy Khimii i Khimicheskoy Technologii,* 2005, № 3, p. 106-112.

[9] Kozlov G.V., Burya A.I., Zaikov G.E. *J. Appl. Polymer Sci.,* 2006, v. 100, № 4, p. 2817-2820.

[10] Burya A.I., Kozlov G.V., Kholodilov O.V. *Vestnik Polotskogo Gosuniversiteta, Seriya* B, 2005, № 6, p. 36-39.

[11] Burya A.I., Kozlov G.V. *Novye Materialy i Technologii v Metallurgii i Mashinostroenii,* 2006, № 2, p. 35-37.

[12] Kozlov G.V., Shetov R.A., Mikitaev A.K. Vysokomolek. *Soed.* A, 1987, v. 29, № 9, p. 2012-2013.

[13] Kozlov G.V., Sanditov D.S. Anharmonic Effects and Physical-Mechanical Properties of Polymers. *Novosibirsk, Nauka,* 1994, 261 p.

[14] Sanditov D.S., Kozlov G.V. *Fizika i Khimiya Stekla,* 1996, v. 22, № 2, p. 97-106.

[15] Kozlov G.V., Shustov G.B., Burya A.I. Izvestiya VUZov, Severo-Kavkazsk. region, technich. *Nauki,* 2004, № 9, p. 156-162.

[16] Kozlov G.V., Lipatov Yu.S. *Mechanika Kompozitnykh Materialov,* 2003, v. 39, № 1, p. 89-96.

[17] Phillips D.C., Harris B. In book: Polymer Engineering Composites. Ed. Richardson M.O.W. London, *Applied Science Publishers LTD,* 1978, p. 50-146.

[18] Burya A.I., Chigvintseva O.P. *Sovremennoe Mashinostroenie,* 1999, № 2, p. 28-32.

[19] Aloev V.Z., Burya A.I., Kozlov G.V. In book: "Selected Works of Republican Scientific Seminar "Mechanics", *Nal'chik, KBSAA,* 2004, p. 61-66.

[20] MacClintock F.A., Argon A.S. Mechanical Behaviour of Materials. *Reading, Addison-Wesley Publishing Company, Inc.,* 1966, 443 p.

[21] Novikov V.U., Kozlov G.V., Lipatov Yu.S. *Plast. Massy,* 2003, № 10, p. 4-8.

[22] Pfeifer P. In book: *Fractals in Physics.* Ed. Pietronero L., Tosatti E. Amsterdam, Oxford, New York, Tokyo, North-Holland, 1986, p. 72-81.

[23] Kozlov G.V., Yanovskiy Yu.G., Lipatov Yu.S. *Mekhanika Kompozitsionnykh Materialov i Konstruktsiy,* 2003, v. 9, № 3, p. 398-448.

[24] Burya A.I., Chigvintseva O.P., Suchilina-Sokolenko S.P. Polyarylates. Synthesis, Properties, Composite Materials. Dnepropetrovsk, *Nauka i Obrazovanie,* 2001, 152 p.

[25] Burya A.I., Kozlov G.V. *Fiziko-Khimicheskaya Mekhanika Polimerov,* 2007, v. 43, № 2, p. 124-126.

[26] Peacock A.J., Mandelkern L. *J. Polymer Sci.: Part B: Polymer Phys.,* 1990, v. 28, № 11, p. 1917-1941.

[27] Serdyuk V.D., Kozlov G.V., Mashukov N.I., Mikitaev A.K. *J. Mater. Sci. Techn.,* 1996, v. 5, № 2, p. 55-60.

[28] Edwards D.C. *J. Mater. Sci.,* 1990, v. 25, № 12, p. 4175-4185.

[29] Kozlov G.V., Novikov V.U. Synergetics and Fractal Analysis of Cross-linked Polymers. Moscow, *Klassika,* 1998, 112 p.

[30] Kalinchev E.L., Sakovtseva M.B. *Properties and Processing of Thermoplastics.* Leningrad, Khimiya, 1983, 288 p.

[31] Burya A.I., Kozlov G.V., *Rula I.V. Works of Third International Conf. "Materials and Coatings in Extreme Conditions: Studies, Application, Utilization".* 13-17 Sept. 2004, Katsiveli, Crimea, Ukraine. Katsiveli, 2004, p. 52-54.

[32] Lipatov Yu.S. *Physical-Chemical Grounds of Polymers Filling.* Moscow, Khimiya, 1991, 259 p.

[33] Kozlov G.V., Zaikov G.E. In book: *"Fractals and Local Order in Polymeric Materials"*. Ed. Kozlov G., Zaikov G. New York, Nova Science Publishers, Inc., 2001, p. 55-63.
[34] Burya A.I., Kozlov G.V., Kazakov M.E. In Works Collection of 2th International Conf. *"Study, Elaboration and Application of High Technologies in Industry"*. V. 3. 30 May-2 June 2005, Sankt-Peterburg, p. 214-216.
[35] Burya A.I., Kozlov G.V., Ignatov M.I. Mater. of II International Sci.-Pract. Conf. *"Dynamics of Scientific Studies-2003"*. V. 17, Dnepropetrovsk, 2003, p. 17.
[36] Yu Z., Ait-Kadi A., Brisson J. *Polymer Engng. Sci.,* 1991, v. 31, № 16, p. 1222-1227.
[37] Aloev V.Z., Kozlov G.V., Zaikov G.E. *Russian Polymer News,* 2001, v. 6, № 4, p. 63-65.
[38] Wunderlich B. *Macromolecular Physics. V. 2. Crystal Nucleation, Growth, Annealing.* New York, San Francisco, London, Academic Press, 1978, 561 p.
[39] Bershtein V.A., Egorov V.M. *Differential Scanning Calorimetry in Physics-Chemistry of Polymers.* Leningrad, Khimiya, 1990, 256 p.
[40] Burya A.I., Aloev V.Z., Kozlov G.V. *Uspekhi Sovremennogo Estestvoznaniya,* 2004, № 11, p. 65.
[41] Burya A.I., Kozlov G.V., Vishnyakov L.R., Pavlenko O.A. *Technologicheskie Sistemy,* 2004, № 5-6, p. 27-31.
[42] Shogenov V.N., Belousov V.N., Potapov V.V., Kozlov G.V., Prut E.V. *Vysokomolek. Soed. A,* 1991, v. 33, № 1, p. 155-160.
[43] Sanditov D.S., Kozlov G.V. *Fizika i Khimiya Stekla,* 1995, v. 21, № 6, p. 547-576.
[44] Kozlov G.V., Zaikov G.E. Structure of the Polymer Amorphous State. *Utrecht-Boston, Brill Academic Publishers,* 2004, 465 p.
[45] Belousov V.N., Kozlov G.V., Mashukov N.I., Lipatov Yu.S. *Doklady AN,* 1993, v. 328, № 6, p. 706-708.
[46] Kozlov G.V., Beloshenko V.A., Gazaev M.A., Novikov V.U. *Mechanika Kompozitnykh Materialov,* 1996, v. 32, № 2, p. 270-278.
[47] Brown N. *Mater. Sci. Engng.*, 1971, v. 8, № 1, p. 69-73.
[48] Aleksanyan G.G., Berlin Al.Al., Gol'danskiy A.V., Grineva N.S., Onishchuk V.A., Shantorovich V.P., Safonov G.P. *Khimicheskaya Fizika,* 1986, v. 5, № 9, p. 1225-1234.
[49] Kozlov G.V., Novikov V.U. *Uspekhi Fizicheskikh Nauk,* 2001, v. 171, № 7, p. 717-764.
[50] Balankin A.S., Bugrimov A.L., Kozlov G.V., Mikitaev A.K., Sanditov D.S. *Doklady AN,* 1992, v. 326, № 3, p. 463-466.
[51] Beloshenko V.A., Kozlov G.V. *Mechanika Kompozitnykh Materialov,* 1994, v. 30, № 4, p. 451-454.
[52] Abaev A.M., Beloshenko V.A., Kozlov G.V., Mikitaev A.K. *Fizika i Technika Vysokikh Davleniy,* 1998, v. 8, № 2, p. 102-109.
[53] Burya A.I., Kozlov G.V., Sviridenok A.I. *Doklady NAN Belarusi,* 2007, v. 51, № 2, p. 100-102.
[54] Balankin A.S., Bugrimov A.L. *Vysokomolek. Soed.* A, 1992, v. 34, № 10, p. 135-139.
[55] Burya A.I., Kozlov G.V., Kiprich V.V. Mater. 7th International Conf. "Research and Development in Mechanical Industry. *RaDMI-2007"*. 20 Sept. 2007, Belgrade, Serbia, p. 749-750.

[56] Burya A.I., Kozlov G.V., Vishnyakov L.R. *Novye Materialy i Technologii,* 2004, № 2, p. 41-44.

[57] Budtov V.P. *Physical Chemistry of Polymer Solutions.* Sankt-Peterburg, Khimiya, 1992, 384 p.

[58] Haward R.N. *Macromolecules,* 1993, v. 26, № 22, p. 5860-5869.

[59] Haward R.N. *J. Polymer Sci.: Part B: Polymer Phys.,* 1995, v. 33, № 8, p. 1481-1494.

[60] Belousov V.N., Kozlov G.V., Mikitaev A.K., Lipatov Yu.S. *Doklady AN SSSR,* 1990, v. 313, № 3, p. 630-633.

[61] Burya A.I., Kozlov G.V., Kazakov M.E., Golovyatinskaya V.V., Murakhovskaya M.L. In *Works Collection of 2th International Conf.* "Study, Elaboration and Application of High Technolodies in Industry". V. 4, Sankt-Peterburg, 7-9 Febr. 2006, p. 87-91.

[62] Bekichev V.I., Bartenev G.M. *Vysokomolek. Soed.* A, 1972, v. 14, № 3, p. 545-550.

[63] Kozlov G.V., Shustov G.B., Zaikov G.E. Izvestiya VUZov, *Severo-Kavkazsk. region, estestv. nauki,* 2003, № 2, p. 58-60.

[64] Flory P.J. Polymer J., 1985, v. 17, № 1, p. 1-12.

[65] Haward R.N., Thackray G. *Proc. Roy. Soc. London,* 1968, v. A302, № 1471, p. 453-472.

[66] Edwards S.F., Vilgis T.A. *Polymer,* 1987, v. 28, № 3, p. 375-378.

[67] Burya A.I., Kozlov G.V., Rula I.V. *J. Guangdong Non-ferrous Metal Sep.,* 2005, v. 15, № 2, 3, p. 216.

[68] Kozlov G.V., Serdyuk V.D., Dolbin I.V. *Materialovedenie,* 2000, № 12, p. 2-5.

[69] Kozlov G.V., Temiraev K.B., Shetov R.A., Mikitaev A.K. *Materialovedenie,* 1999, № 2, p. 34-39.

[70] Solomko V.P. *Filled Crystallizing Polymers.* Kiev, Naukova Dumka, 1980, 264 p.

[71] Burya A.I., Kozlov G.V. *Synergetics and Fractal Analysis of Polymer Composites, Filled with Short Fibers.* Dnepropetrovsk, Porogi, 2008, 258 p.

[72] Narisawa I. *Strength of Polymeric Materials.* Moscow, Khimiya, 1987, 400 p.

[73] Gent A.N., Madan S. *J. Polymer Sci.: Part B: Polymer Phys., 1*989, v. 27, № 7, p. 1529-1542.

[74] Kozlov G.V., Sanditov D.S., Serdyuk V.D. *Vysokomolek. Soed.* B, 1993, v. 35, № 12, p. 2067-2069.

[75] Kaush H.H. *Polymer Fracture.* Berlin, Heidelberg, New York, Springer-Verlag, 1978, 432 p.

[76] Burya A.I., Kozlov G.V., Aloev V.Z. *Mater. of II All-Russian Sci.-Pract. Conf. "New Polymer Composite Materials".* Nal'chik, KBSU, 2005, p. 171-176.

[77] Kozlov G.V., Burya A.I., Zaikov G.E. In book: Polymers, Polymer Blends, Polymer Composites and Filled Polymers. Synthesis, Properties, Application. Ed. Mikitaev A., Ligidov M., Zaikov G. New York, *Nova Science Publishers,* Inc., 2006, p. 22-29.

[78] Bokshitskiy M.N. *Prolonged Strength of Polymers.* Moscow, Khimiya, 1978, 308 p.

[79] Askadskiy A.A. *Deformation of Polymers.* Moscow, Khimiya, 1973, 448 p.

[80] Obert L. In book: Fracture. *An Advanced Treatise.* V. 7, Ch. 1. Ed. Liebowitz H. New York, London, Academic Press, 1972, p. 59-128.

[81] Mosolov A.B., Borodich F.M. *Doklady AN,* 1992, v. 324, № 3, p. 546-549.

[82] Mosolov A.B. *Zhurnal Technicheskoy Fiziki,* 1991, v. 61, № 7, p. 57-60.

[83] Kozlov G.V., Beloshenko V.A., Shogenov V.N., Lipatov Yu.S. *Doklady NAN Ukraine,* 1995, № 5, p. 100-102.

[84] Burya A.I., Kozlov G.V., Chigvintseva O.P., Chaika L.V. *Scientific Works of National Technical University (Donetsk),* 2008, № 95, p. 102-107.

[85] Gol'dstein R.V., Mosolov A.B. *Doklady AN SSSR,* 1991, v. 319, № 4, p. 840-844.

[86] Kozlov G.V., Burya A.I., Ovcharenko E.N. In book: *The Seventh Region: Science and Practice.* Ed. Kanchukoev V. Nal'chik, Polygraphservice and T, 2006, p. 72-76.

[87] Leidner J., Woodhams R.T. *J. Appl. Polymer Sci.,* 1974, v. 18, № 8, p. 1639-1654.

[88] Novikov V.U., Kozlov G.V. *Plast. Massy,* 2004, № 8, p. 12-23.

[89] Yamamoto I., Miyata H., Kobayashi I. *"Benibana" International Symposium.* Absracts, Yamagata, Japan, 1990, p. 184-189.

[90] Kozlov G.V., Novikov V.U. *Prikladnaya Fizika,* 1997, № 1, p. 77-84.

[91] Burya A.I., Kozlov G.V., Chigvintseva O.P. *Fiziko-Khimicheskaya Mekhanika Materialov,* 2005, v. 41, № 3, p. 61-65.

[92] Izotov A.D., Balankin A.S., Lazarev V.B. *Neorganicheskie Materialy,* 1993, v. 29, № 7, p. 883-893.

[93] Henkee C.S., Kramer E.J. *J. Polymer Sci.: Polymer Phys.* Ed., 1984, v. 22, № 4, p. 721-737.

[94] Wu S. *J. Polymer Sci.: Part B: Polymer Phys.,* 1989, v. 27, № 4, p. 723-741.

[95] Aharoni S.M. *Macromolecules,* 1985, v. 18, № 12, p. 2624-2630.

[96] Aharoni S.M. *Macromolecules,* 1983, v. 16, № 9, p. 1722-1728.

[97] Burya A.I., Kozlov G.V. *Polimernui Zhurnal,* 2004, v. 26, № 4, p. 244-248.

[98] Kozlov G.V., Kolodei V.S., Lipatov Yu.S. *Materialovedenie,* 2002, № 11, p. 34-39.

[99] Zhenyi M., Langford S.C., Dickinson J.T., Engelhard M.H., Baer D.R. *J.Mater. Res.,* 1991, v. 6, № 1, p. 183-195.

[100] McCauley J.L. Int. *J. Modern Phys.* B, 1989, v. 3, № 6, p. 821-852.

[101] Novikov V.U., Kozlov G.V. *Materialovedenie, 1*999, № 12, p. 8-14.

[102] Kozlov G.V., Zaikov G.E. *The Structural Stabilization of Polymers: Fractal Models.* Utrecht-Boston, Brill Academic Publishers, 2006, 345 p.

[103] Ivanova V.S., Kuzeev I.R., Zakirnichnaya M.M. Synergetics and Fractals. *Universality of Materials Mechanical Behaviour.* Ufa, Publishers USNTU, 1998, 366 p.

[104] Burya A.I., Kozlov G.V., Chigvintseva O.P. *Mater. of Sci.-Techn. Conf. "New and Nontraditional Technologies of Resource- and Energy-Saving",* 2-4 June 2004, Odessa. Kiev, 2004, p. 23-24.

[105] Folmanis G.E. *Works of International Interdisciplinary Symposium "Fractals and Applied Synergetics,* FiPS-03". Moscow, Publishers MSOU, 2003, p. 303-308.

[106] Kuzeev I.R., Samigullin G.Kh., Kulikov D.V., Zakirnichnaya M.M. Complex Systems in Nature and Engineering. Ufa, *Publishers USNTU,* 1997, 225 p.

[107] Bucknall C.B. Toughened Plastics. London, *Applied Science,* 1977, 318 p.

Chapter 4

THERMOPHYSICAL PROPERTIES OF POLYMER COMPOSITES

4.1. THERMAL CONDUCTIVITY OF COMPOSITES FILLED WITH SHORT FIBERS

For polymer composites thermal conductivity description the approach, based on assumption, that these materials can be considered as resistors systems, is often used. Such approach is universal for any conductivity phenomenon [1]. The assumption of complete geometrical order in phases distribution is general and inevitable at theoretical analysis of conductivity phenomena in composite solid mediums. It is supposed that fibers are distributed in polymeric matrix uniformly on the same distance and parallel to each other. However, real composite materials, prepared as a result of technological operation entire complex fulfillment, have structure, differing essentially from the ideal model. Besides, insufficient knowledge of fibrous fillers themselves and matrix properties imposes in its turn restrictions on theoretical equations application possibilities for thermophysical properties prediction of composite materials [1]. Therefore the bulk conductivity coefficient of fibers system, is often used for composites thermal conductivity description, which takes into account not physical properties only, but also geometrical features of composite material. The authors [2] used the considered approach for carbon plastics on the basis of phenylone thermal conductivity description.

In Figure 4.1 the dependence of thermal conductivity coefficient λ_T on components blending duration in rotating electromagnetic field t for carbon plastics on the basis of phenylone is adduced. As it follows from the adduced plot, at small $t \leq 30$ s the values λ_T are relatively small and at $t \geq 60$ s the higher values λ_T are observed and the dependence $\lambda_T(t)$ reaches asymptotic branch (plateau). Let us remind, that the dependences of the main structural characteristics (for example, structure fractal dimension d_f, Figure 1.1) on t for the considered carbon plastics have typically synergetic character: within the range t=5-120 s the approximately sigmoid dependence $d_f(t)$ is observed, which at $t \geq 120$ s approaches constant value d_f. Such dependence type is specific for periodical (quasiperiodical) structures with subsequent transition of system to chaotic behaviour [3]. Hence, the comparison of Figure 1.1

data and the dependence $\lambda_T(t)$, adduced in Figure 4.1, supposes that the most high thermal conductivity coefficient for the studied carbon plastics is achieved at uniform (chaotic) fibers distribution in polymeric matrix [2].

The value λ_T can be theoretically calculated according to the known Debay equation [4]:

$$\lambda_T = \frac{1}{3} C \overline{\vartheta} \overline{l}_r , \qquad (4.1)$$

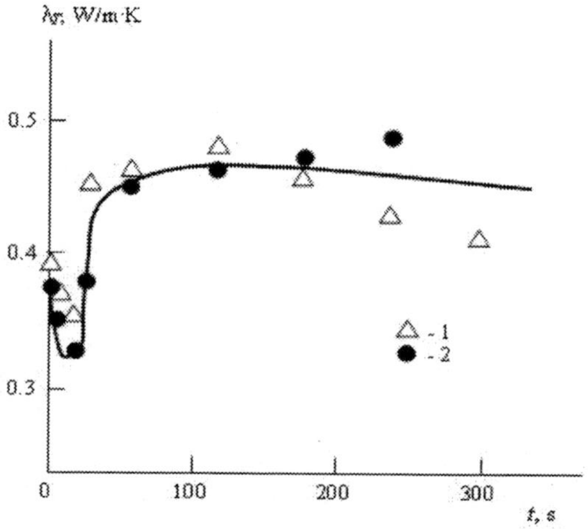

Figure 4.1. The dependence of thermal conductivity coefficient λ_T on components blending duration in rotating electromagnetic field t for carbon plastics on the basis of phenylone, prepared with magnetic (1) and mechanical (2) separation application [2].

where C is heat capacity, $\overline{\vartheta}$ is sound average rate in material, $\overline{l}_r$ is average length of phonons free run.

The value $\overline{\vartheta}$ can be determined according to the formula [5]:

$$\overline{\vartheta} = \left(\frac{E}{\rho} \right)^{1/2} , \qquad (4.2)$$

where E is elasticity modulus, ρ is material density.

As $\overline{l}_r$ for amorphous polymers structure heterogeneity characteristic size is accepted. As it was shown in paper [6], local order domain length is such size and in paper [7] the indicated size is accepted equal to polymer segment length. It is significant, that within the frameworks of the cluster model of polymers amorphous state structure these two $\overline{l}_r$ definitions are coincided [8]. The indicated model postulates that local order domain (cluster) consists of several collinear statistical segments with length l_{st}, determined according to the equation (1.14).

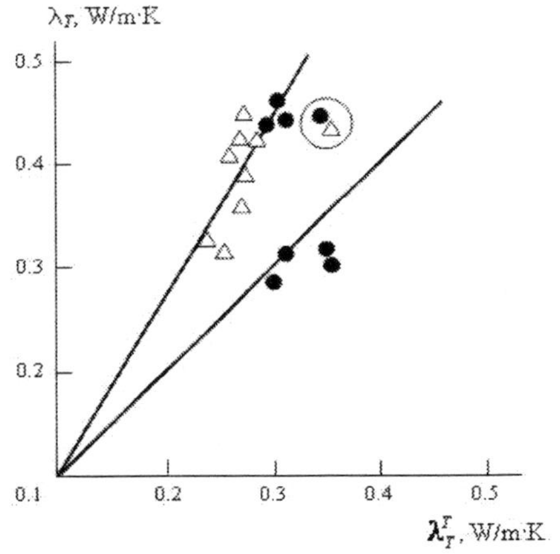

Figure 4.2. The comparison of experimental λ_T and calculated according to the equation (4.1) λ_T^T thermal conductivity coefficient values for carbon plastics on the basis of phenylone, prepared with magnetic (1) and mechanical (2) separation application. The data for samples with structure negative feedback are ringed by a small circle (at t=60 s) [2].

In Figure 4.2 the comparison of λ_T experimentally obtained and calculated according to the equation (4.1) λ_T^T thermal conductivity coefficient values for the studied carbon plastics is adduced. As one can see, this correlation breaks up into two straight lines, and one of them corresponds to λ_T and λ_T^T equality. On this straight line four data points are set, which have the smallest λ_T values (Figure 4.1). Let us remind that the equation (4.1) was obtained for one-phase materials. This supposes that the smallest λ_T values are defined only by polymeric matrix structure. Two data points (ringed by a small circle) are obtained for t=60 s and are related to structures with negative feedback, i.e. they are transitional from quasiperiodic structure to chaotic one [9]. The condition $\lambda_T > \lambda_T^T$ is fulfilled for the remaining composites, including all carbon plastics with chaotic distribution of fibers in polymeric matrix. In other words, in the given case filler fibers totality increases carbon plastics thermal conductivity. To account this effect the authors [2] used the known method [1]: in the equation for conductivity coefficient calculation instead of real physical magnitudes, characterizing composite material separate components, and instead of variable parameters, coordinating experimental and calculated data, coefficient of bulk conductivity can be included, which accounts not only physical properties, but also composite material geometrical features. Such approach is due to the fact, that fibrous and porous insulating materials are usually characterized by their bulk properties just so [1].

The equation for composites thermal conductivity description can be written as follows [1]:

$$\ln\lambda_T = \varphi_f \ln\lambda_T^* + \left(1-\varphi_f\right)\ln\lambda_T^m, \qquad (4.3)$$

where λ_T^* is effective bulk thermal conductivity coefficient of fibers system, λ_T^m is thermal conductivity coefficient of polymeric matrix.

Accepting λ_T^m=0.30 w/m·K according to the data of Figure 4.2, the value λ_T^* can be calculated from the equation (4.3). In Figure 4.3 the dependence of λ_T^* on carbon plastics structure governing parameter – fiber orientation factor η, which characterizes fibers system state in polymeric matrix [10], is adduced. As one can see, again by analogy with the data of Figure 4.2, the linear dependence $\lambda_T^*(\eta)$ is observed, which is broken up into two straight lines and between them the data points for carbon plastics with t=60 s, possessing negative feedback in the structure, are discovered. The value λ_T^* for carbon plastics, prepared at small t, is much smaller than for samples with fibers chaotic distribution in polymeric matrix, i.e. prepared at $t\geq120$ s. This circumstance defines λ_T growth for carbon plastics, for which components blending duration in rotating electromagnetic field is large enough.

Hence, the adduced above results have shown that thermal conductivity coefficient of carbon plastics on the basis of phenylone is defined by two constituents: polymeric matrix thermal conductivity and filler fibers totality (system) thermal conductivity. The first constituent depends on matrix structure and is well described by Debay equation. The second constituent depends on fibers distribution in polymeric matrix and reaches the largest value at uniform (chaotic) distribution of the indicated fibers.

The considered above method of composites thermal conductivity description (the equation (4.3)) is not sole. Within the frameworks of fractal analysis for filler particles (aggregates of particles) system description fractal dimension of filler particles network D_n is used, which characterizes polymeric matrix space filling density by filler particles or fibers [11, 12]. Further the fractal model considers components A and B random mixture, which has good and bad conductive sections [13]. This model corresponds completely to polymer composites, thermal conductivity coefficients of carbon fibers and polymeric matrix of which can differ on three orders [1]. Two limiting cases of this problem deserve particular attention [13]:

1) random network of resistors (RNR). In this case it is supposed that sections, occupied by bad conductor B, have a zero conductivity;
2) random superconducting network (RSN). In this case conductivity of good conductor is infinite.

The authors of papers [14, 15] studied thermal conductivity of carbon plastics on the basis of phenylone within the frameworks of the considered above fractal model.

As it was shown in paper [13], thermal conductivity coefficient λ_T for the two considered above limiting cases is given by the following relationships:

$$\lambda_T \sim L^{d_u} \quad (4.4)$$

for RSN and

$$\lambda_T \sim L^{d_w - D_n} \quad (4.5)$$

for RNR.

In the relations (4.4) and (4.5) L is cluster size, d_u is a fractal dimension of its nonscreening perimeter, d_w is a dimension of walk on fractal.

The dimensions necessary for subsequent calculations can be calculated as follows. d_u value is determined according to the equation [16]:

$$d_u = (D_n - 1) + \left(\frac{d - D_n}{d_w} \right), \qquad (4.6)$$

where d is the dimension of Euclidean space, in which fractal is considered (it is obvious, in our case d=3) and the dimension d_w can be estimated according to Alexander-Orbach rule [17]:

$$d_w = \frac{3}{2} D_n . \qquad (4.7)$$

And at last, the dimension D_n for the studied carbon plastics can be calculated with the aid of the equation (1.31).

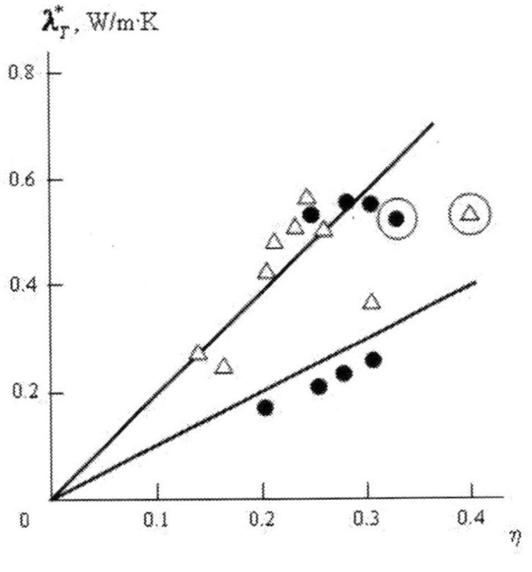

Figure 4.3. The dependence of fibers system thermal conductivity coefficient λ_T^* on fibers orientation factor η for carbon plastics on the basis of phenylone. The designations are the same as in Figure 4.2 [2].

In Figure 4.4 the dependence of λ_T on parameter L^{d_u} is adduced, where the value L is arbitrarily accepted equal to 5 relative units. As one can see, in composite simulation case by RSN half of the data points lies on a straight line, passing through coordinates origin (that is not shown in Figure), i.e. they correspond to the indicated model and anther half of the points discovers large scatter, i.e. it does not correspond to RSN model. Actually, the adduced in Figure 4.5 dependence $\lambda_T(L^{\xi})$ confirms this supposition, where ξ is conductivity exponent, which is equal to d_u in RSN case and to $(d_w$-$D_n)$ in RNR case. The obtained data are approximated by one straight line for both pointed out models, which passes through coordinates origin. Data certain scatter for the dependence $\lambda_T(L^{\xi})$ can be due, as a minimum, by two factors: L arbitrary choice and the condition L=const and by d_w approximate estimation according to Alexander-Orbach rule as well. Another possible cause of the indicated scatter can be polymeric matrix thermal conductivity variations.

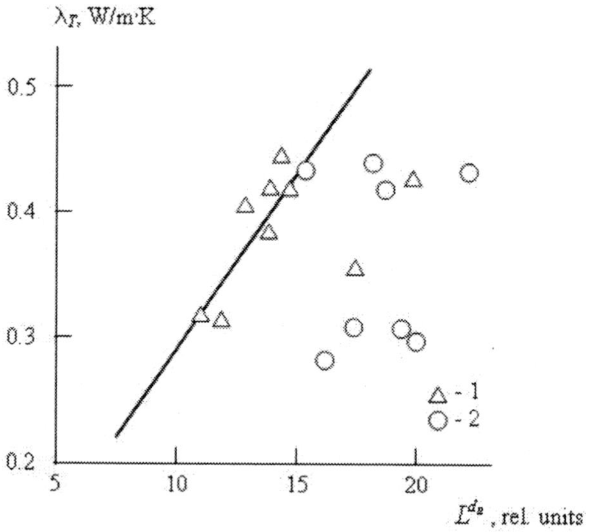

Figure 4.4. The dependence of thermal conductivity coefficient λ_T on parameter L^{d_u} for carbon plastics on the basis of phenylone, prepared with magnetic (1) and mechanical (2) separation application [14].

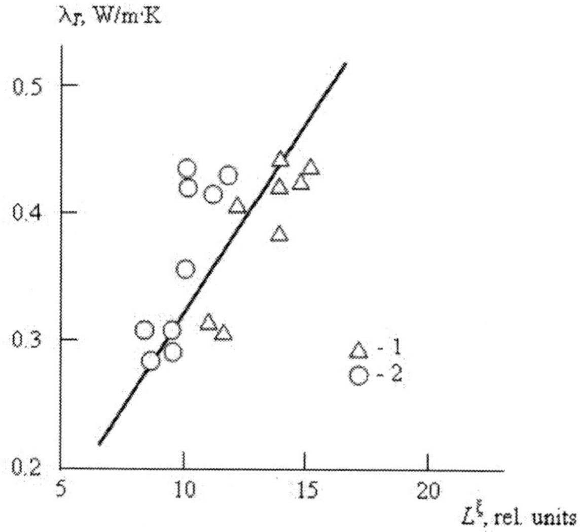

Figure 4.5. The dependence of thermal conductivity coefficient λ_T on parameter L^{ξ} at $\xi=d_u$ (1) and $\xi=(d_w-D_n)$ (2) for carbon plastics on the basis of phenylone [14].

The transition from one composites thermal conductivity model to the other occurs at $D_n \approx 2.62$: at $D_n < 2.62$ RSN model is correct and at $D_n > 2.62$ – RNR model. Let us note, that the value D_n is connected with the governing parameter of carbon plastics synergetic structure – fibers orientation factor η - by the relationship (1.32). Therefore, from this relationship it follows, that RSN model is correct for $\eta < 0.313$ and RNR model – for $\eta > 0.313$ [15].

Let us consider physical premises of the observed transition from RSN model to RNR one. In RSN limit heat transfer has no geometrical restrictions and can be realized in both polymeric matrix and fibers network (system). Therefore the value λ_T is controlled by dimension d_u or by a number of accessible for this process sites of filler fibers network [13]. In RNR case heat transfer in regions with zero conductivity, i.e. in polymeric matrix, is impossible and λ_T value is controlled by dimension D_n, i.e. by fibers network dimension. In Figure 4.6 the dependence $\lambda_T(D_n)$ is adduced, which breaks down into two linear sections, boundary of which is dimension $D_n \approx 2.62$ (the vertical shaded line in Figure 4.6). For $D_n < 2.62$ (RSN limit) the dependence $\lambda_T(D_n)$ is approximated as follows [15]:

$$\lambda_T = 0.90(D_n - 2), \text{ w/m·K}, \qquad (4.8)$$

and for $D_n > 2.62$ (RNR limit) the dependence $\lambda_T(D_n)$ approximation has the following form [15]:

$$\lambda_T = 0.51(D_n - 2), \text{ w/m·K}. \qquad (4.9)$$

Hence, in RSN limit the faster growth λT at Dn increase is observed, than in RNR limit. Such conclusion follows immediately from the relationships (4.4) and (4.5) comparison, since for the considered carbon plastics du>(dw-Dn) [15].

Hence, the stated above results have shown that thermal conductivity coefficient of carbon plastics on the basis of phenylone can be described within the frameworks of fractal model. Depending on filler fibers network (system) dimension such description can be obtained by the application of two limiting cases: random network of resistors (RNR) and random superconducting network (RSN) [14, 15].

The model within the frameworks of percolation theory corresponds to each from the indicated above limiting cases (RNR and RSN) [13]. The authors [18] studied thermal conductivity of carbon plastics on the basis of phenylone within the frameworks of random mixtures conductivity percolation conception.

As it has been shown in chapter 3, the change of components blending duration in rotating electromagnetic field t results to elasticity modulus E_c of the studied carbon plastics variation (E_c=2.13-3.33 GPa). Within the frameworks of percolation theory the authors [19] proposed the equation (1.76) for the value E_c determination. As it has been noted above [20] (see section 1.3 as well), this equation in the original form is inapplicable for the description of E_c behaviour of carbon plastics, since for them φ_f=const, but the value E_c changes in more than 1.5 times. However, the equation (1.76) can be used for the calculation of filling effective volume degree φ_f^{ef}, which reflects the filler fibers system structure, defining composite macroscopic properties (in the given case E_c). Within the frameworks of percolation theory parameter φ_f^{ef} is an equivalent to effective fraction of solid-state component p in random mixture, i.e. $\varphi_f^{ef}=p$. Besides, the comparison of this parameter with fractal dimension of filler particles network D_n is of interest. The similar comparison is adduced in Figure 1.65. The correlation of structural characteristics of fibers system (network) in carbon plastics on the basis of phenylone is adduced in Figure 4.7. As one can see, p linear growth at D_n increase is observed. Since D_n changes within the limits of $2.0 \le D_n < 3.0$ [11], then this gives the possibility to estimate values p variation within the range of 0.060-0.264. The correlation between $p=\varphi_f^{ef}$ and D_n analytically is expressed by the equation (1.77).

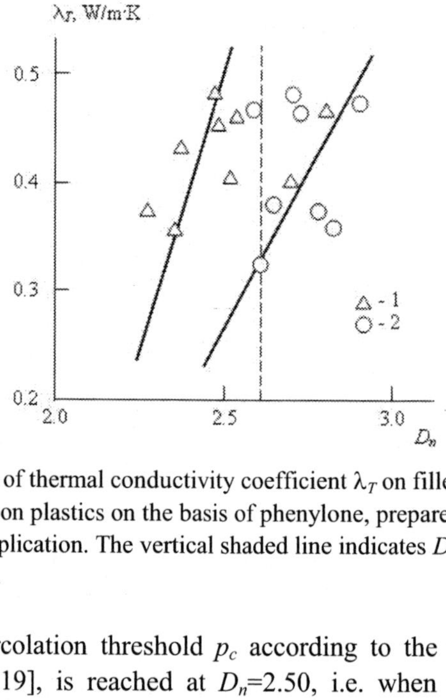

Figure 4.6. The dependences of thermal conductivity coefficient λ_T on filler fibers network (system) fractal dimension D_n for carbon plastics on the basis of phenylone, prepared with magnetic (1) and mechanical (2) separation application. The vertical shaded line indicates D_n value, bordering for RSN and RNR limiting cases [14].

Let us note, that percolation threshold p_c according to the touching spheres scheme, which is equal to 0.16 [19], is reached at D_n=2.50, i.e. when fibers network dimension becomes equal to Witten-Sander cluster dimension [21]. The physical significance of the adduced in Figure 4.7 correlation is that one. The fibers strong anisotropy (the average ratio of length to diameter is approximately equal to 300) results to the fact, that p value is defined to a considerable extent by their orientation. The authors [18] considered the value ρ_s in the direction parallel to sample axis and fibers orientation factor η increasing in this direction meant actually filling effective degree enhancement in that direction and corresponding reduction – in perpendicular one. The plot of Figure 4.8, gives the confirmation to this supposition where the dependence $p(\eta)$ is adduced. As one can see, the linear p growth at η increase is obtained. Besides, the plot of Figure 4.8 shows that fibers certain orientation ($\eta \approx 0.15$, compare with the equation (3.11)) already exists at nominal value $p=\varphi_f$.

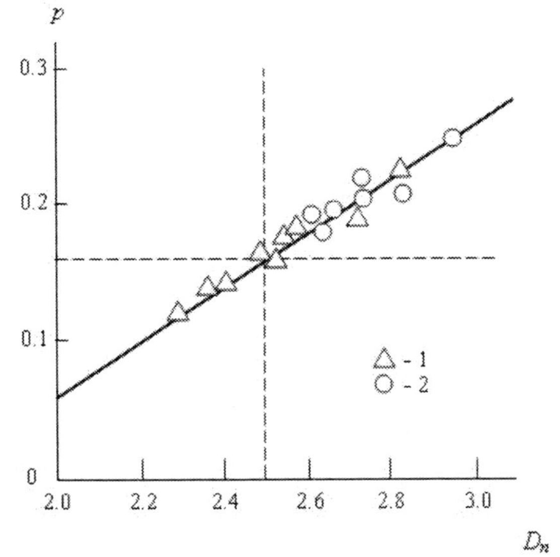

Figure 4.7. The dependence of solid-state component concentration p on fibers network fractal dimension D_n for carbon plastics on the basis of phenylone, prepared with magnetic (1) and mechanical (2) separation application. The horizontal shaded line points out percolation threshold p_c, the vertical one – the Witten-Sander cluster dimension [18].

Thermal conductivity λ_T within the frameworks of RSN limiting case can be described by the following percolation relationship [13]:

$$\lambda_T \sim (p_c - p)^{-s}, \qquad (4.10)$$

for $p<p_c$ and in RNR limit – by the relationship [13]:

$$\lambda_T \sim (p - p_c)^{\mu} \qquad (4.11)$$

for $p>p_c$.

The exponents in the relationships (4.10) and (4.11) are determined as follows. s value is equal to [13]:

$$s = [d_u - (d-2)]\nu_p, \qquad (4.12)$$

where du is dimension of nonscreening (accessible for heat transfer) fibers network perimeter, d is dimension of Euclidean space, in which a fractal is considered, νp is percolation index, equal to 0.88 [22].

It is obvious, that in the considered case d=3, dimension d_u is determined according to the equation (4.6) and d_w – according to Alexander-Orbach rule (the equation (4.7)). Another variant of d_w estimation is adduced in paper [13]:

$$d_w = d - d_u. \qquad (4.13)$$

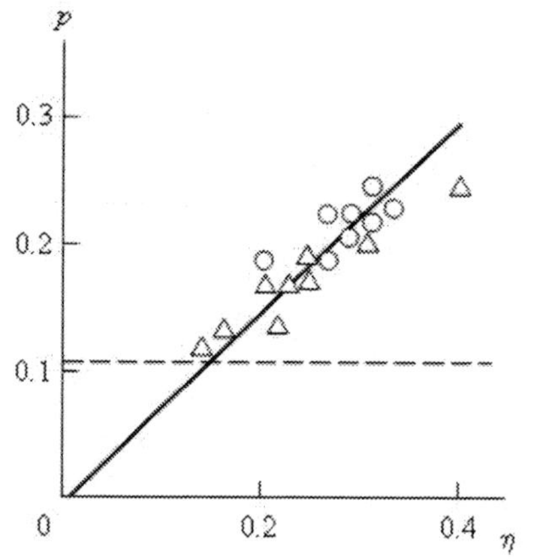

Figure 4.8. The dependence of solid-state component concentration p on fibers orientation factor η for carbon plastics on the basis of phenylone. The horizontal shaded line points out nominal value φ_f. The designations are the same as in Figure 4.7 [18].

The exponent μ in the relationship (4.11) is determined according to the following equation [13]:

$$\mu = [(d_w - D_n) - (d - 2)]\nu_p \,. \tag{4.14}$$

In Figure 4.9 the comparison of experimental λ_T and calculated in relative units according to the relationships (4.10) and (4.11) λ_T^T thermal conductivity coefficient values for RSN and RNR limiting cases is adduced. As one can see, in both cases λ_T value corresponds well enough to theoretical estimation according to the indicated relationships, i.e. the linear correlation, passing through coordinates origin, is observed.

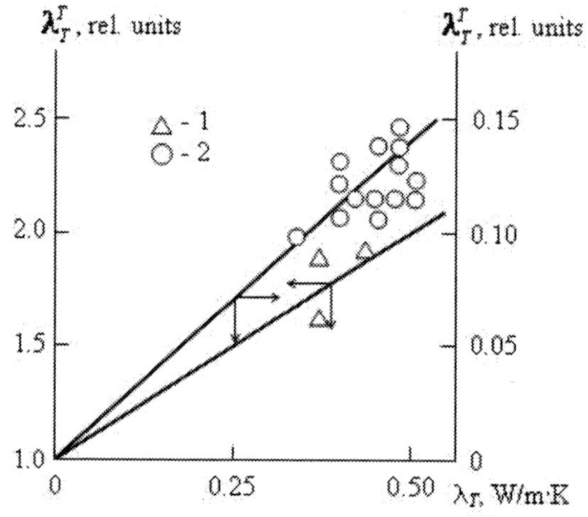

Figure 4.9. The comparison of experimental λ_T and calculated according to the relationships (4.10) (1) and (4.11) (2) λ_T^T thermal conductivity coefficient values for carbon plastics on the basis of phenylone [18].

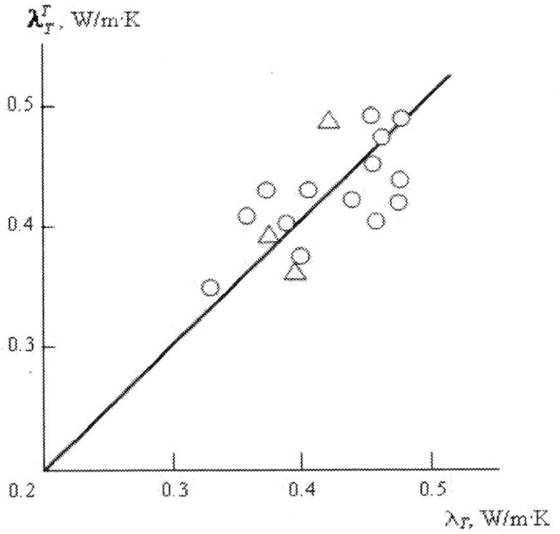

Figure 4.10. The comparison of experimental λ_T and calculated theoretically (explanations are in text) λ_T^T thermal conductivity coefficient values for carbon plastics on the basis of phenylone. The designations are the same as in Figure 4.9 [18].

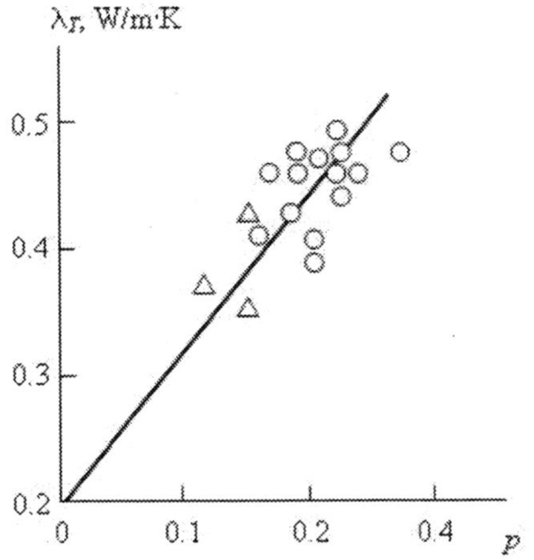

Figure 4.11. The dependence of thermal conductivity coefficient λ_T on solid-state component concentration p for carbon plastics on the basis of phenylone. The designations are the same as in Figure 4.9 [18].

To calculate the value λ_T^T in the same units, in which experimental values λ_T were determined, the constant coefficients for the relationships (4.10) and (4.11) were obtained, equal to 0.25 and 3.42 v/m·K, accordingly. The adduced in Figure 4.10 comparison of calculated thus values λ_T^T and λ_T shows their good correspondence – the average discrepancy makes up less than 10 %. Taking into account, that experimental determination λ_T precision makes up 4-6 % [6], then the offered methodic can be used for prediction of thermal conductivity coefficient of carbon plastics.

Let us consider in conclusion physical nature of heat transfer mechanisms, which correspond to RSN and RNR limiting cases. In the RSN case ($p<p_c$) heat transfer through filler fibers is realized instantly, since they are considered superconductors, and therefore the value λ_T is limited by heat transfer through polymeric matrix sections, dividing fibers. Since polymeric matrix filling density is defined by the dimension D_n, then the higher D_n (and p), the smaller these sections and the larger λ_T. In RNR limit ($p>p_c$) heat transfer is realized through fibers contact sites and is defined again by D_n value, since d_u increase at D_n growth follows from the equation (4.6). In Figure 4.11 the dependence $\lambda_T(p)$ for the considered carbon plastics is adduced. As one can see, in general case λ_T value in the RNR case is higher than in RSN case, but at percolation threshold ($p=p_c=0.16$) these values are approximately the same. The dependence $\lambda_T(D_n)$ will have a similar shape in virtue of linear correlation between p and D_n (Figure 4.7).

Hence, the considered above results have shown that the filling effective degree, equivalent to solid-state component concentration in percolation theory, changes at the same nominal contents of filler, depending on composite preparation conditions. Percolation model in both limiting cases (RSN and RNR) can be used for the description and prediction of carbon plastics thermal conductivity coefficient.

4.2. Heat Expansion of Composites Filled with Short Fibers

Heat expansion is one of the criteria of engineering polymer composites choice for their usage in that or an other quality [23]. As a rule, polymers have high heat expansion coefficient, that makes it difficult to use them in contact with other materials. One of the methods of polymers heat expansion decrease is the introduction of fillers in to them. This method allows to reduce heat expansion coefficient up to five times. The authors [24] fulfilled the study of heat expansion for composites on the basis of polyarylate (PAr), filled with four types of short fibers (terlon, vniivlon, uglen, which are organic fibers, and glass fiber) with their contents 5-35 mass. % [25].

In Figure 4.12 three solid curves 1-3 give three main types of the dependences of bulk thermal expansion coefficient γ_c on filler fibers volume contents φ_f. The straight line 1 illustrates the case, when adhesion is absent between composite two phases and at $\gamma_m>\gamma_f$ (indices "*m*" and "*f*" are related to polymeric matrix and filler, accordingly, and index "*c*" – to composite) and the absence in interfacial layers of compression residual strain polymeric matrix at heating will be independently expanded on filler and in this case $\gamma_c=\gamma_m$ [23]. The straight line 2 corresponds to simple mixtures rule:

$$\gamma_c = \gamma_m(1-\varphi_f)+\gamma_f\varphi_f. \qquad (4.15)$$

This rule is true for the ideal case only, when each phase expands independently from each other.

And at last, the curve 3 corresponds to Terner equation [23]:

$$\gamma_c = \frac{\gamma_m(1-\varphi_f)K_m + \gamma_f\varphi_f K_f}{(1-\varphi_f)K_m + \varphi_f K_f}, \tag{4.16}$$

where K is bulk elasticity modulus.

At the derivation of the equation (4.16) the strain equality method was used for the calculation of the thermal expansion coefficient γ of mixtures, proceeding from the density, elasticity modulus, heat expansion coefficient and mass relation of the constituent components. If the made assumptions are correct, then the formula (4.16) should be applicable to anisotropic composite materials as well [23].

Besides, experimentally determined values of γ_c for the four indicated above types of composites on the basis of PAr are shown in Figure 4.12. As one can see, the data for glassy plastics correspond to theoretical straight line 2, i.e. to a simple mixtures rule. As it was indicated above, that meant PAr and glassy fiber independent expansion. The data for the three studied organoplastics correspond precisely enough (within the limits of 15 %) to the theoretical curve, calculated according to Terner formula (the equation (4.16)). The indicated correspondence allows to make the important conclusion – for glassy plastics adhesion level on interfacial boundary polymer-filler is low, whereas for organoplastics one should expect very strong adhesion [23].

The similar dependences $\gamma_c(\varphi_f)$ were observed earlier too. So, the dependence $\gamma_c(\varphi_f)$, similar to adduced in Figure 4.12 for glassy plastics, was obtained for composites on the basis of curing nonsaturated polyether filled with short glass fibers with length of 6.3 mm [23]. The dependences $\gamma_c(\varphi_f)$, well described by Terner formula, were obtained for composite on the basis of the same polyether, but filled with continuous glassy fiber [23]. It is significant, that in both indicated cases composites samples with both short and continuous fiber were anisotropic. The considered above data allow to explain the dependences $\gamma_c(\varphi_f)$ behaviour for composites on the basis of PAr as follows. First, the dependences $\gamma_c(\varphi_f)$ behaviour similarity for the considered composites and anisotropic samples, which consists in the absence of the dependences, intermediate between curves 2 and 3 (Figure 4.12), supposes high orientation degree of fibers in composites on the basis of PAr in virtue of their preparation using technology with components preliminary blending in rotating electromagnetic field [23]. Secondly, as the study of composites on the basis of PAr structure by IR-spectroscopy methods showed, strong chemical bonds were formed between organic fibers (uglen, vniivlon, terlon) surface and polymeric matrix, which were absent in the case of glass fiber using [25]. This means weak adhesion for PAr-glass fiber system and possibility of components independent expansion [23]. For organoplastics chemical bonds availability defines strong adhesion on the interfacial boundary, owing to which the value γ_c approaches rapidly to γ_f at φ_f growth. It is obvious, that in considered aspect organic fibers as filler are more preferable: as it follows from the data of Figure 4.12, that now at $\varphi_f \approx 0.3$ the value γ_c is smaller in about 5 times than the corresponding parameter for PAr.

If to suppose, that the value γ_c for composites on the basis of PAr is defined by interfacial adhesion level, then the correlation between this parameter and fibers surface total area follows, which will be proportional to $\varphi_f^{2/3}$. This assumption confirms the adduced in Figure 4.13 dependence $\gamma_c(\varphi_f^{2/3})$, which is linear and extrapolates to the value γ for PAr (which is equal to 24×10^{-5} K^{-1}) at φ_f=0. Let us note as well, that the data of Figures 4.12 and 4.13 suppose the adhesion level reduction for glassy plastics at φ_f growth: at φ_f<0.1 the dependence $\gamma_c(\varphi_f)$ for the indicated composites corresponds to Terner equation and at φ_f>0.1 to simple mixtures rule, that appeals to adhesion essential deterioration in the last case [24].

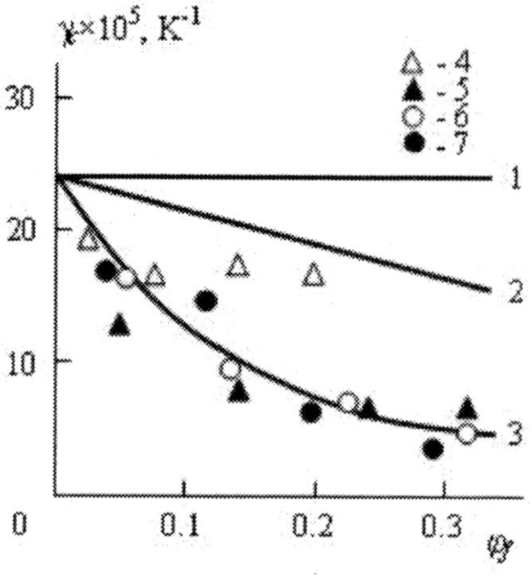

Figure 4.12. The dependences of volume heat expansion coefficient γ_c on fibers volume contents φ_f for composites on the basis of PAr. The theoretical curves: adhesion absence on the interfacial boundary (1), mixtures rule (2), Terner equation (3). The experimental data for fillers: glassy fiber (4), vniivlon (5), terlon (6) and uglen (7) [24].

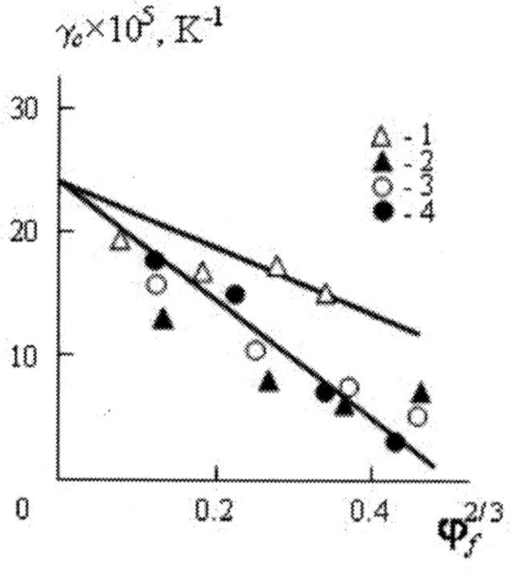

Figure 4.13. The dependences of volume thermal expansion coefficient γ_c on parameter $\varphi_f^{2/3}$ value for composites on the basis of PAr filled with glassy fiber (1), vniivlon (2), terlon (3) and uglen (4) [24].

Therefore, the adduced above data have demonstrated that, as minimum, two factors influence on heat expansion coefficient value of composites filled by short fibers: fibers orientation degree and interfacial adhesion level on boundary polymer-filler at the same volume contents of the latter. An organic fibers usage for heat expansion coefficient reduction gives much stronger effect in comparison with glassy fiber owing to chemical bonds formation on interfacial boundary.

REFERENCES

[1] Ziblend H. In book: Polymer Engineering Composites. Ed. Richardson M.O.W. London, *Applied Science Publishers LTD,* 1978, p. 284-319.

[2] Kozlov G.V., Burya A.I., Ovcharenko E.N. *Izwestiya KBNC RAN,* 2006, № 1, p. 137-141.

[3] Ivanova V.S., Kuzeev I.R., Zakirnichnaya M.M. Synergetics and Fractals. Universality of Materials Mechanical Behaviour. Ufa, *Publishers USNTU,* 1998, 366 p.

[4] Kittel C. Introduction in Solid Body Physics. Moscow, *Nauka,* 1978, 696 p.

[5] Krasil'nikov V.A., Krylov V.V. Introduction in Physical Acoustics. Moscow, *Nauka,* 1984, 400 p.

[6] Perepechko I.I., Golub' P.D., Nasonov A.D. *Vysokomolek. Soed.* B, 1984, v. 26, № 6, p. 470-473.

[7] Zabashta Yu.F. *Vysokomolek. Soed. B,* 1991, v. 33, № 1, p. 42-46.

[8] Kozlov G.V., Zaikov G.E. Structure of the Polymer Amorphous State. Utrecht-Boston, *Brill Academic Publishers,* 2004, 465 p.

[9] Kozlov G.V., Burya A.I., Zaikov G.E. *J. Appl. Polymer Sci.,* 2006, v. 100, № 4, p. 2817-2820.

[10] Burya A.I., Kozlov G.V., Kholodilov O.V. *Vestnik Polotskogo Universiteta, seriya B,* 2005, № 6, p. 36-39.

[11] Kozlov G.V., Mikitaev A.K. *Mekhanika Kompozitsionnykh Materialov i Konstruktsiy,* 1996, v. 2, № 3-4, p. 144-157.

[12] Novikov V.U., Kozlov G.V. *Mekhanika Kompozitnykh Materialov,* 1999, v. 35, № 3, p. 269-290.

[13] Stanley H.E. In book: *Fractals in Physics.* Ed. Pietronero L., Tosatti E. Amsterdam, Oxford, New York, Tokyo, North-Holland, 1986, p. 463-477.

[14] Kozlov G.V., Burya A.I., Dolbin I.V. *J. Engng. Thermophysics,* 2005, v. 13, № 2, p. 129-135.

[15] Kozlov G.V., Burya A.I., Dolbin I.V., Zaikov G.E. *J. Appl. Polymer Sci.,* 2006, v. 100, № 5, p. 3828-3831.

[16] Meakin P., Coniglio A., Stanley H.E., Witten T.A. *Phys. Rev.* A, 1986, v. 34, № 4, p. 3325-3340.

[17] Alexander S., Orbach R. *J. Phys. Lett. (Paris),* 1982, v. 43, № 17, p. L625-L631.

[18] Burya A.I., Kozlov G.V., Vishnyakov L.R., Rula I.V. Works of 4th International Conf. "Theory and Practice of Products Manufacture Technologies from Composite Materials and New Metal Alloys (TPCMM). *Corporative Nano- and CALS-Technologies in Industry Science-Capacious Branches".* 26-29 April 2005, Moscow, 2005, p. 67-72.

[19] Bobryshev A.N., Kozomazov V.N., Babin L.O., Solomatov V.I. Synergetics of Composite Materials. Lipetsk, *NPO ORIUS,* 1994, 154 p.

[20] Kozlov G.V., Burya A.I., Zaikov G.E. In book: Molecular and High Molecular Chemistry: Theory and Practice. Ed. Monakov Yu., Zaikov G. New York, *Nova Science Publishers, Inc.*, 2006, p. 131-137.

[21] Witten T.A., Sander L.M. *Phys. Rev. Lett.,* 1981, v. 47, № 19, p. 1400-1403.

[22] Sokolov I.M. *Uspekhi Fizicheskikh Nauk,* 1986, v. 150, № 2, p. 221-256.

[23] Holliday L., Robinson J.D. In book: Polymer Engineering Composites. Ed. Richardson M.O.W. London, *Applied Science Publishers LTD,* 1978, p. 241-283.

[24] Kozlov G.V., Burya A.I., Dolbin I.V. *Voprosy Materialovedeniya,* 2005, № 3(43), p. 51-54.

[25] Burya A.I., Chigvintseva O.P., Suchilina-Sokolenko S.P. Polyarylates. Synthesis, Properties, Composite Materials. Dnepropetrovsk, *Nauka i Obrazovanie,* 2001, 152 p.

[26] Fomichev A.I., Burya A.I., Gubenkov M.G. *Elektronnaya Obrabotka Materialov,* 1978, № 4, p. 26-27.

Chapter 5

THERMAL STABILITY OF POLYMER COMPOSITES: THE DESCRIPTION WITHIN THE FRAMEWORKS OF STRANGE (ANOMALOUS) DIFFUSION MODEL

As it is known [1], an intensive thermal degradation temperature T_d characterizes thermal stability of polymers. As thermal stability characteristic according to [2] "the limiting temperature is accepted, at which a polymer chemical change reflecting on its properties takes place". Thermal stability is defined with the help of the thermogravimetric analysis (TGA). Further the temperature determined according to intersection of the tangents to thermogravimetric curve two branches will be accepted as T_d [1].

The parameter T_d importance predetermines vast enough literature, dedicated to the study of the value T_d dependence on polymer characteristics [1-5]. Nevertheless, these studies take into account T_d interrelation only with polymer chemical constitution by one way or another: the existence of "weak links" [3], of either groups in polymer chain [4], polymer chain defects [5] etc. However, taken into account physical structure of polymeric material in all these studies is not, although the authors [6] showed, that for PC films, obtained from different solvents (that assumes polymer chemical constitution invariability) T_d difference could reach 120 K. This result assumes a strong influence of polymer structure at testing temperature on its thermal stability. In this sense the most suitable for studing objects are polymer films of one and the same polymer, obtained from different solvents, and polymer composites. For these classes of polymeric materials large enough variations of their physical structure without polymer (polymeric matrix) chemical constitution change are possible. Therefore the authors [7] made the study of the polymeric materials structure influence on their thermal stability and obtained the analytic interrelation of T_d and structural characteristics on the example of PC films, obtained from solutions in five different solvents, and carbon plastics on the basis of phenylone. The fractal analysis methods were used as studies theoretical basis.

As a rule, the temperature T_d is higher than the temperature of the so-called transition "liquid 1 – liquid 2" T_{ll}. At T_{ll} there is transition of polymeric melt from "liquid with the fixed structure" (where the residual structural ordering is observed) to the true liquid state or "the structureless liquid". Nevertheless "the absence of structure" of a melt refers to the absence of supermolecular structure, but the macromolecular coil structure in a melt remains the important structural factor (in essence, the only structural factor at $T>T_{ll}$) [9]. The value T_{ll} is connected with glass transition temperature of polymer T_g by the following simple relationship [10]:

$$T_{ll} \approx (1.20 \pm 0.05)T_g . \qquad (5.1)$$

The structure of a macromolecular coil, which is a fractal object [11], can be most precisely characterized with the help of its fractal (Hausdorff) dimension Δ_f. This dimension is a true structural characteristic of macromolecular coil, since it characterizes its elements distribution in space [12].

As it was noted above, within the frameworks of strange (anomalous) diffusion on fractal objects two its main types can be selected: slow and fast diffusion. In the basis of this division lies the dependence of mobile reagent displacement s on time t (see the equation (1.57)), where for the classical case β=1/2, for the slow diffusion β<1/2 and for the fast one - β>1/2 [8].

The authors [13] showed within the frameworks of fractional derivatives theory the interrelation of Δ_f and β, which was analytically expressed as follows:

$$\beta = \frac{\Delta_f - 1}{4} \qquad (5.2)$$

for the slow diffusion and

$$\beta = \frac{\Delta_f - 1}{\Delta_f} \qquad (5.3)$$

for the fast one. The structural boundary between the indicated kinds of diffusion should be considered the value Δ_f=2.5 at general variation $2.0 \le \Delta_f < 3.0$: at Δ_f<2.5 (less compact macromolecular coils) the oxidizer (oxygen) fast diffusion is realized, at Δ_f>2.5 – the slow one [7].

The following equation is used for the value T_d theoretical estimation [7]:

$$\Delta_f = c(T_d - T_g)^{\beta}, \qquad (5.4)$$

where c is constant, which is accepted equal to 0.093 for the fast diffusion and to 0.305 – for the slow one [6].

The equation (5.4) defines three factors influencing on polymeric materials thermal stability: polymer chemical constitution, characterized by its glass transition temperature T_g, polymer melt structure, characterized by the dimension Δ_f and the type (intensity) of oxidizer diffusion, connected with structure and characterized by the exponent β [7].

As the estimations of the constant c according to best fitting method of theory and experiment have shown, this constant has different values for various polymers and different diffusion types. So, in case of carbon plastics on the basis of phenylone c=0.20 for the fast diffusion and 0.50 for the slow one and in PC case (see above) c=0.093 and 0.305, accordingly. The comparison of experimental T_d and calculated according to the equation (5.4) T_d^T with using of the indicated above values c intensive degradation temperature for two polymeric materials in Figure 5.1 is adduced. As one can see, a good correspondence of theory and experiment (the average discrepancy of T_d and T_d^T values makes up ~ 2 %) [7].

Analysis of constant c values in the equation (5.4) shows their increase at polymer T_g growth (taking into account the data for glassy plastics on the basis of PAr [7]). This circumstance allows to plot the dependence $c(T_g)$ for the case of fast diffusion, which is shown in Figure 5.2. This dependence can be written analytically as follows [7]:

$$c = 1.25 \times 10^{-3}\left(T_g - 350\text{K}\right). \tag{5.5}$$

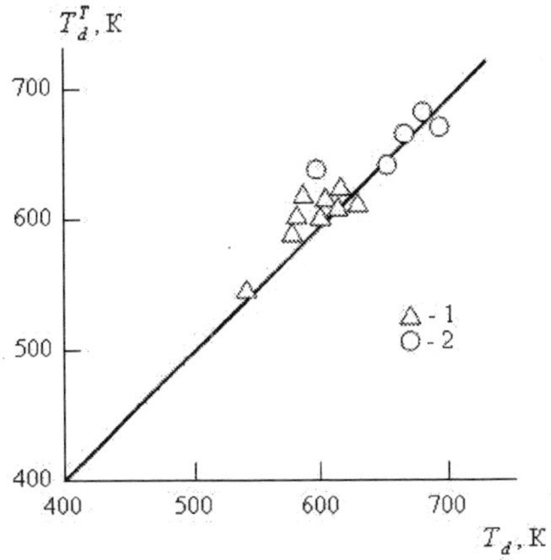

Figure 5.1. The comparison of experimental T_d and calculated according to the equation (5.4) T_d^T intensive degradation temperature values for carbon plastics on the basis of phenylone (1) and PC films (2) [7].

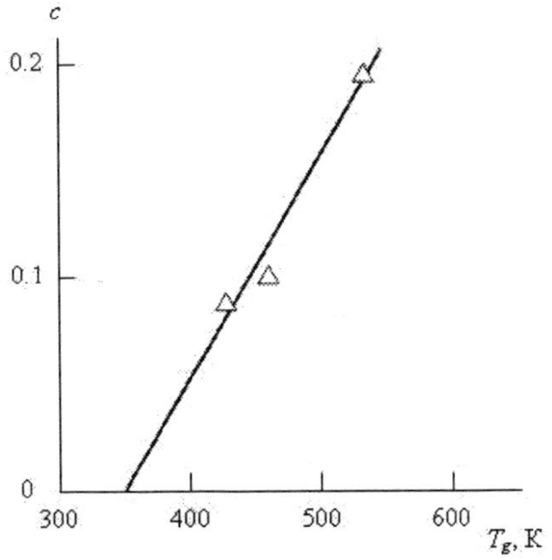

Figure 5.2. The dependence of coefficient c in the equation (5.4) on glass transition temperature T_g [7].

The equation (5.5) allows to make a number of conclusions. Firstly, it is true for polymers with relative high glass transition temperature (T_g>350 K). Secondly, T_g increase results to c growth and according to the equation (5.4), to structure role in T_d definition reduction. Thirdly, the last circumstance explains larger values c for the slow diffusion in comparison with the last one for the same polymer: structural factor role in the slow diffusion case is essentially lower, that can be supposed a priori. With the equation (5.5) accounting the formula (5.4) could be rewritten as follows [7]:

$$\Delta_f = 1.25\times10^{-3}\left(T_g - 350\text{K}\right)\left(T_d - T_g\right)^{\beta}, \qquad (5.6)$$

that allows to avoid different values of coefficient c for various polymers and diffusion types.

In Figure 5.3 the comparison of experimental T_d and calculated according to the equation (5.6) T_d^T intensive degradation temperature values for PC films, carbon plastics on the basis of phenylone and glassy plastics on the basis of PAr is adduced. As one can see, the equation (5.6) gives precise enough estimation of T_d value (an average discrepancy of T_d and T_d^T makes up ~ 2.8 %) within T_g range ~ 160 K [7].

And at last, in Figure 5.4 the dependence of ratio (T_d/T_g) on T_g, calculated according to the equation (5.6), is shown. As follows from the plot of Figure 5.4, the ratio (T_d/T_g) goes rapidly to unity or $T_d \rightarrow T_g$ at T_g raising and at $T_g \approx 700$ K distinction between T_d and T_g is in the limits of the calculation error, i.e. <3 %. Just such relation between T_d and T_g at T_g raising is experimentally observed [1].

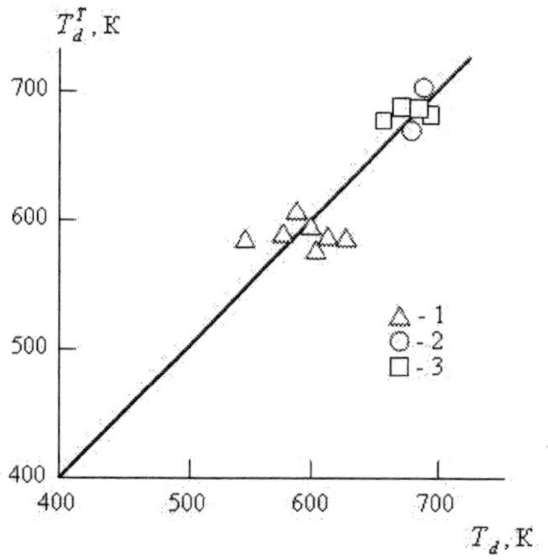

Figure 5.3. The comparison of experimental T_d and calculated according to the equation (5.6) T_d^T intensive degradation temperature values for carbon plastics on the basis of phenylone (1), PC films (2) and glassy plastics on the basis of PAr (3) [7].

Therefore, the stated above results have shown the important role of polymer melt structure in the intensive degradation temperature definition or polymeric materials thermal stability. The last characteristic depends on polymers chemical constitution, expressed through glass transition temperature, macromolecular coil compactness in melt and type of oxidizer diffusion, which is also defined by a coil structure. The intensive degradation temperature calculation according to the offered equations shows a good correspondence to experiment. The structure role in thermal stability definition is reduced at glass transition temperature increasing [7].

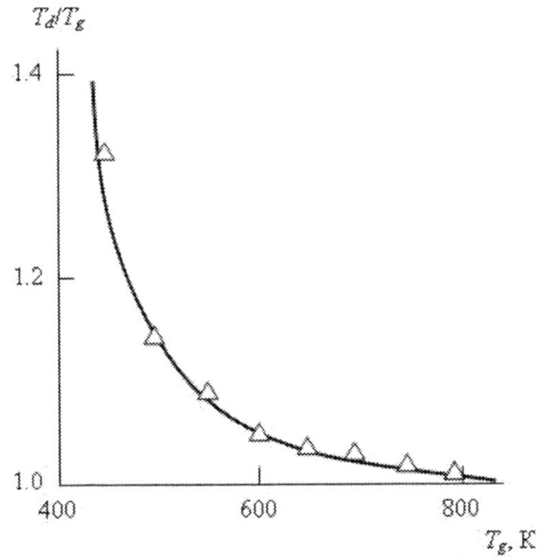

Figure 5.4. The dependence of the ratio T_d/T_g on glass transition temperature T_g calculated according to the equation (5.6). Vertical shaded line indicates the condition $T_d \approx T_g$ [7].

As it is known, the temperature, corresponding to either percent sample mass loss in thermal degradation process, is often used as temperature T_d. In papers [14, 15] the temperature of 20 % sample mass loss $T_{20\%}$, determined by TGA method is used as such indicator at thermal stability studing of composites on the basis of PAr filled with organic fibers of three types – uglen, terlon and vniivlon [16]. The calculation of theoretical values $T_{20\%}$ ($T^T_{20\%}$) was made according to the equation (5.4), where constant c was accepted equal to 0.093. Since the value Δ_f for the studied composites is relatively small (Δ_f=2.39-2.50, see table 5.1), then the exponent β was determined for the fast diffusion case (the equation (5.3)).

In Table 5.1 the comparison of 20 % sample mass loss temperature values $T_{20\%}$ and $T^T_{20\%}$ for the indicated composites is adduced. As one can see, a good correspondence of theory and experiment (the average discrepancy Δ makes up ~ 2.7 %) is obtained. All the obtained values $T^T_{20\%}$ fall in the range of experimentally determined temperatures of thermooxidative degradation proceeding – 673-763 K [16].

It is significant that for these organoplastics the calculation $T^T_{20\%}$ for a slow diffusion case has also given a good correspondence with the experiment. It is obvious, that it is due to the values Δ_f closeness to boundary value Δ_f=2.5 (table 5.1). Nevertheless, the calculation error in the last case is somewhat larger [17].

Therefore, the obtained for the composites on the basis of PAr results showed the dependence of $T_{20\%}$ on melt structure, characterized by the dimension Δ_f. T_g increasing and Δ_f decreasing results to $T_{20\%}$ growth [15].

Table 5.1. The comparison of experimental and theoretical TGA characteristics for organoplastics on the basis of PAr [15]

Reinforcing fiber		$T_{20\%}$, K	T_g, K	Δ_f	$T^T_{20\%}$, K	β	Δ, %
Type	Contents, mass %						
Uglen	5	713	453	2.47	693	0.595	2.9
	15	718	458	2.45	711	0.592	1.0
	25	723	463	2.42	733	0.587	1.4
	35	728	473	2.39	755	0.582	3.6
Terlon	5	713	463	2.45	716	0.592	0.4
	15	718	463	2.43	731	0.588	1.8
	25	723	473	2.40	764	0.583	5.4
	35	728	488	2.40	764	0.583	4.7
Vniivlon	5	703	459	2.50	691	0.600	1.7
	15	708	461	2.50	693	0.600	2.1
	25	710	463	2.50	695	0.600	2.1
	35	713	473	2.50	705	0.600	1.1

REFERENCES

[1] Askadskiy A.A. *Structure and Properties of Thermostable Polymers.* Moscow, Khimiya, 1981, 320 p.

[2] Korshak V.V. *Chemical Constitution and Temperature Characteristics of Polymers.* Moscow, Nauka, 1970, 419 p.

[3] Van Krevelen D.W. Properties and Chemical Structure of Polymers. Amsterdam, London, New York, *Elsevier Publishing Company,* 1972, 414 p.

[4] Sazanov Yu.N., Kudryavtsev V.V., Svetlichnyi V.N., Fedorova G.N., Antonova T.A., Aleksandrova E.P. *Vysokomolek. Soed.* A, 1983, v. 25, № 5, p. 975-978.

[5] Mikitaev A.K., Beriketov A.S., Korshak V.V., Taova A.Zh. *Vysokomolek. Soed.* A, 1983, v. 25, № 8, p. 1691-1696.

[6] Dolbin I.V., Kozlov G.V., Bazheva R.Ch., Shustov G.B. *Mater. of V All-Russian Sci.-Pract. Conf. "New Chemical Technologies: Production and Application".* Penza, PSU, 2003, p. 42-45.

[7] Dolbin I.V., Burya A.I., Kozlov G.V. *Teplofizika Vysokikh Temperatur,* 2007, v. 45, № 3, p. 355-358.

[8] Zelenyi L.M., Milovanov A.V. *Uspekhi Fizicheskikh Nauk,* 2004, v. 174, № 8, p. 809-852.

[9] Kozlov G.V., Shustov G.B., Zaikov G.E. *J. Appl. Polymer Sci.,* 2004, v. 93, № 5, p. 2343-2347.

[10] Berstein V.A., Egorov V.M. *Differential Scanning Calorimetry in Physics-Chemistry of Polymers.* Leningrad, Khimiya, 1990, 256 p.

[11] Vilgis T.A. *Physica A,* 1988, v. 154, № 2, p. 341-354.

[12] Ivanova V.S., Kuzeev I.R., Zakirnichnaya M.M. Synergetics and Fractals. Universality of Materials Mechanical Behaviour. Ufa, *Publishers USNTU,* 1998, 366 p.

[13] Shogenov V.Kh., Akhkubekov A.A., Akhkubekov R.A. Izvestiya VUZov, Severo-Kavkazsk. *region, estestv. nauki,* 2004, № 1, p. 46-50.

[14] Dolbin I.V., Burya A.I., Kozlov G.V., Ol'khovaya G.G. *Mater. IV All-Russian Sci.-Pract. Conf. "Innovations in Mechanical Engineering".* Penza, PSU, 2004, p. 45-48.

[15] Dolbin I.V., Burya A.I., Kozlov G.V. *Fundamental'nye Issledovaniya,* 2005, № 3, p. 39-41.

[16] Burya A.I., Chigvintseva O.P., Suchilina-Sokolenko S.P. Polyarylates. Synthesis, Properties, Composite Materials. Dnepropetrovsk, *Nauka i Obrazovanie,* 2001, 152 p.

[17] Kozlov G.V., Zaikov G.E. The structural Stabilization of Polymers: Fractal Models. Utrecht-Boston, *Brill Academic Piblishers,* 2006, 345 p.

Chapter 6

FRICTION MECHANISMS AND FRICTIONAL WEAR OF POLYMER COMPOSITES

6.1. STRUCTURAL ASPECTS OF FRICTION AND WEAR PROCESSES OF POLYMER COMPOSITES

Polymer composites, filled with short fibers, are perspective materials for application in friction units. These composites are used successfully for traditional engineering and antifriction materials replacement [1].

However, the quantitative structural model absence for polymers until quite recently was an obstacle for these materials friction and frictional wear processes analytical description. The available experimental data [2] unequivocally indicate on polymeric matrix important role in the mentioned processes. This role increases appreciably at more hard regimes (pressure P and sliding rate ϑ) of friction because of temperature raising in a contact zone. The cluster model of polymers amorphous state structure [3, 4] appearance allowed to elucidate the main structural aspects of friction and frictional wear processes. The authors [5] considered this question on the example of carbon plastics on the basis of phenylone.

Let us consider the available grounds of composite polymeric matrix structure and friction and frictional wear main parameters interconnection. Firstly, as paper [1] results have shown, the linear wear intensity I sharp increase (in about 4.5 times) is observed at temperature enhancement T_c in the contact zone from 327 K up to 452 K, i.e. on 125 K. Such behaviour is specific for polymers at testing temperature approaching to glass transition temperature T_g, but not for carbon fibers, whose thermal stability is substantially higher. Secondly, at tests conditions hardness increase, particularly, at specific load p enhancement, there observes fibers orientation in friction direction, their homogeneous distribution violation occurs, a great number of long fibers, un broken in wear process, but pulled out of the matrix appear on a contrabody, that serves as the evidence of its softening on contact spots [2]. The wear products availability on the contrabody indicates just on composite polymeric matrix wear.

In Figure 6.1 the dependence of T_c on the product ($p\vartheta$) for carbon plastic on the basis of phenylone is adduced. As one can see, the monotonous T_c increase at ($p\vartheta$) growth is observed. Significant enhancement of T_c rate reduction at ($p\vartheta$) increase, is a typical sign of process elapsing with fractal object participation [6]. It has been expected, since polymeric matrix structure is a fractal [7]. It is also important to note parameters p and ϑ equivalency in composite surface heating intensification. For linearization of the dependence $T_c(p\vartheta)$ one can use a square root for argument and then the dependence adduced in Figure 6.1 can be analytically written as follows [5]:

$$T_c = 293 + 72(p\vartheta)^{1/2}, \text{ K}, \tag{6.1}$$

where p is given in MPa and ϑ - in m/s.

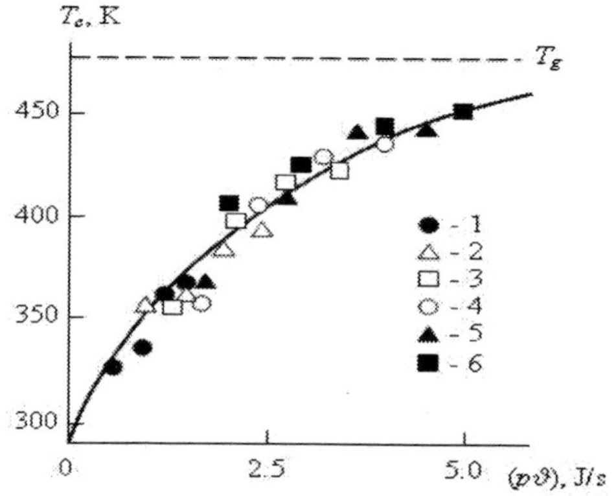

Figure 6.1. The dependence of temperature in contact zone T_c on the value $(p\vartheta)$ for carbon plastics on the basis of phenylone at specific loads p: 0.6 (1), 1.0 (2), 1.4 (3), 1.6 (4), 1.8 (5) and 2.0 MPa (6) [5].

The cluster model [3, 4] assumes that polymers amorphous state structure represents local order domains (clusters), consisted of several collinear densely-packed segments of different macromolecules (amorphous analogue of crystallite with stretched chains) and surrounded by loosely-packed matrix. Since clusters have thermofluctuational nature, then T_c increasing in friction zone should provoke their partial decay and, as consequence, their relative fraction φ_{cl} reduction at the corresponding enhancement of loosely-packed matrix relative fraction $\varphi_{l.m.}$, since the values φ_{cl} and $\varphi_{l.m.}$ are connected with each other by the relationship [3]:

$$\varphi_{l.m.} = 1 - \varphi_{cl}. \tag{6.2}$$

The value φ_{cl} as a function of temperature T_c in composite surface layer can be determined according to the percolation relationship (1.72). Assuming for phenylone T_g=525 K [5], the value φ_{cl} can be calculated according to the equation (1.72) and then $\varphi_{l.m.}$ can be determined according to the equation (6.2). It was shown earlier [1] that $(p\vartheta)$ and, consequently, T_c increase results to friction coefficient f reduction from 0.17 up to 0.07. This assumes possible interconnection of structural characteristics, for example $\varphi_{l.m.}$, and f. In Figure 6.2 the dependence of f on $\varphi_{l.m.}$ reciprocal value is shown. As it follows from the data of this Figure, despite the essential scatter, the tendency of f linear increase at $\varphi_{l.m.}^{-1}$ growth or $\varphi_{l.m.}$ decrease is observed. This correlation can be analytically described as follows [5]:

$$f = 0.05 + 0.138\left(\varphi_{l.m.}^{-1} - 1\right). \qquad (6.3)$$

As the paper [1] data showed, temperature increase in the contact zone resulted to linear wear I enhancement of composite samples. Since T_c enhancement also defines $\varphi_{l.m.}$ growth, then one should expect the correlation between I and $\varphi_{l.m.}$. As the data of Figure 6.3 showed, such correlation actually existed and was analytically expressed as follows [5]:

$$I = 10^{-6}\varphi_{l.m.}^{2}, \text{ m}. \qquad (6.4)$$

The quadratic form of correlation between I and $\varphi_{l.m.}$ confirms the parameter strong influence of the latter on composites linear wear.

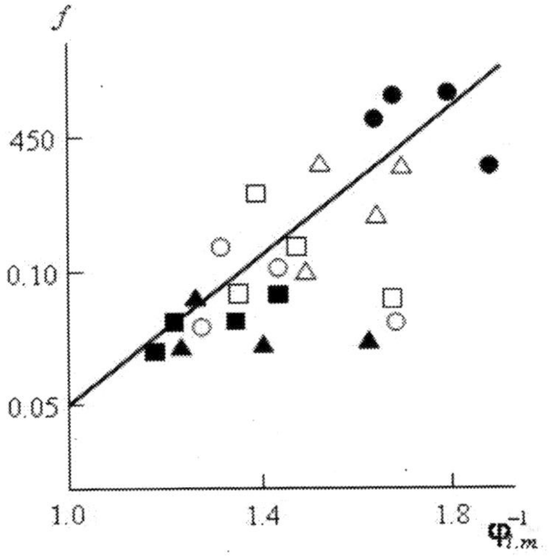

Figure 6.2. The dependence of friction coefficient f on reciprocal value of loosely-packed matrix relative fraction $\varphi_{l.m.}$ for carbon plastics on the basis of phenylone. The designations are the same as in Figure 6.1 [5].

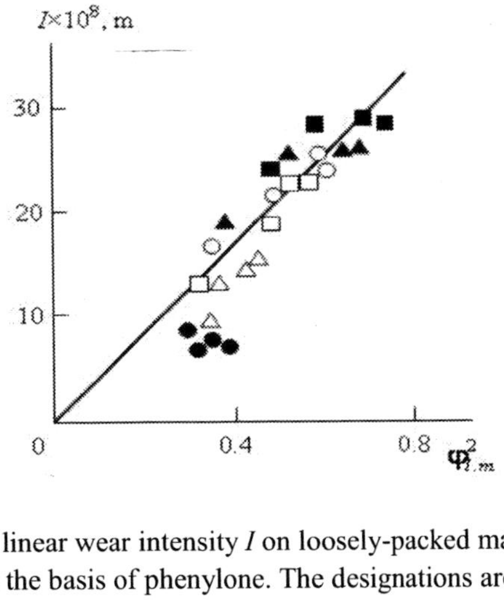

Figure 6.3. The dependence of linear wear intensity I on loosely-packed matrix relative fraction $\varphi_{l.m.}$ quadrate for carbon plastics on the basis of phenylone. The designations are the same as in Figure 6.1 [5].

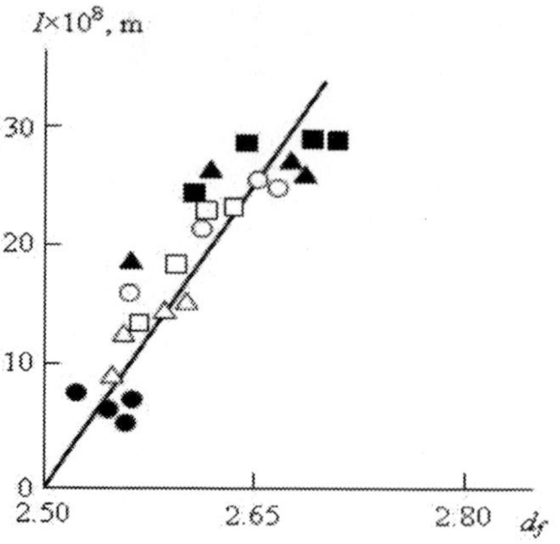

Figure 6.4. The dependence of linear wear intensity I on structure fractal dimension d_f for carbon plastics on the basis of phenylone. The designations are the same as in Figure 6.1 [5].

The stated above results allow to offer the following methodics of friction and wear processes parameters prediction or material choice for using in friction units. This methodics consists in using as the initial parameter of the product ($p\vartheta$), then according to which the equation (6.1) T_c was calculated, further according to the equations (1.72) and (6.2) the value $\varphi_{l.m.}$ is estimated and at last according to the equations (6.3) and (6.4) the operating characteristics f and I are determined. Let us note, that the last ones can be chosen within the necessary limits by selection of polymeric material with required T_g [5].

Let us consider in conclusion a some aspects of fractal analysis application for the mentioned above tasks decision. In Figure 6.4 the dependence of linear wear intensity I on structure fractal dimension d_f for carbon plastics on the basis of phenylone is adduced, which proves to be linear and is described by the empirical equation (1.15).

The precise boundary values of fractal analysis, i.e. fractal dimensions, are one of its undoubtful merits. So, although polymers structure dimension in general case is varied within the limits $2 \le d_f < 3$ [8], the dependence $I(d_f)$ at $I=0$ is extrapolated not to $d_f=2.0$, but to $d_f=2.5$. This does not mean that at $d_f<2.5$ wear at friction will not be observed, since this extrapolation has other physical grounds. The condition $d_f \le 2.50$ perfectly characterizes brittle fracture, the condition $2.50<d_f<2.67$ – quasibrittle (quasiductile) fracture and $d_f \ge 2.67$ – ductile fracture [9]. Therefore, the plot of Figure 6.4 demonstrates that frictional wear of carbon plastics on the basis of phenylone has either quasibrittle or ductile character, the last one is realized at $T_c \ge 433$ K only, that according to the equation (6.1) corresponds to $(p\vartheta)=3.8$ MJ/s.

As it was noted above, from the data of Figure 6.1 the enhancement T_c rate deceleration at ($p\vartheta$) growth follows. This effect has clearly defined physical ground as well, being described within the frameworks of fractal analysis or, more precisely, within the frameworks of fractional integration and differentiation. At influx of mechanical energy to solid, i.e. at its deformation, a heat flow is formed, which is due to deformation (in our case – wearing). The first law of thermodynamics

$$dU = dQ + dW \tag{6.5}$$

ascertains that the internal energy change dU in sample is equal to the sum of work dW, made over sample, and the heat flow dQ in sample. This relationship is true for any deformation, reversible or irreversible. Two thermodynamically irreversible cases exist, for which dQ and dW are equal by absolute value and opposite by sign: uniaxial deformation of Newtonian liquid and ideal elastoplastic deformation. For amorphous glassy polymers deformation has essentially different character: the ratio $Q/W\neq1$ and varies within the limits of 0.36-0.75 depending on testing conditions. In other words, for these materials ideal plasticity is not realized. This effect cause is thermodynamic nonequilibrium of these polymers structure [10].

Lately the mathematical calculus of fractional integration and differentiation has been used for fractal objects description, which is amorphous glassy polymers structure [11]. It was shown [12], that Cantor's set fractal dimension coincided with integral fractional exponent ν_{fr}, which indicated fraction of system states, preserved during its evolution all the time. As it is known [12], Cantor's multitude ("dust") is considered in one-dimensional Euclidean space (d=1) and therefore its fractal dimension d_f obeys condition d_f<1. This supposes that for the fractals, which are considered in Euclidean spaces with d>1 (d=2, 3, …) as fractional exponent ν_{fr} the fractal dimension fractional part must be accepted or

$$\nu_{fr} = d_f - (d-1). \tag{6.6}$$

In this case the value ν_{fr} characterizes system (polymer's structure) states part, preserved during its evolution time [12]. Besides, a very simple interconnection of latent energy U is found and ν_{fr}, is expressed as follows [13]:

$$U = \nu_{fr}. \tag{6.7}$$

The latent energy U is defined as a relative value according to the formula [10]:

$$U = \frac{W-Q}{W}. \tag{6.8}$$

Combination of the equations (6.6)-(6.8) allows to obtain the following relationship, defining heat flow Q in to sample at its wearing [5]:

$$Q = (p\vartheta)(1-\vartheta_{fr}) = (p\vartheta)(d-d_f), \tag{6.9}$$

where in the considered case d=3.

In Figure 6.5 the dependence of T_c on the reduced value $(p\vartheta)(d-d_f)$ is shown, which is linear and extrapolates to the testing temperature T=293 K at $(p\vartheta)$=0. This correlation is analytically expressed as follows [5]:

$$T_c = 293 + 113(p\vartheta)(3-d_f), \text{ K}. \tag{6.10}$$

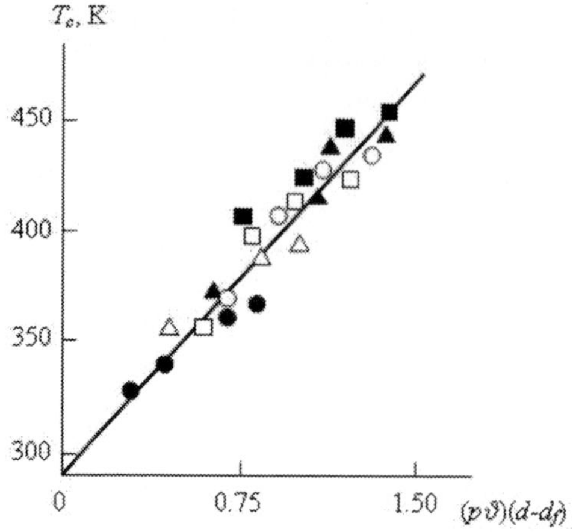

Figure 6.5. The dependence of temperature in the contact zone T_c on the reduced value $(p\vartheta)(d-d_f)$ for carbon plastics on the basis of phenylone. The designations are the same as in Figure 6.1 [5].

It is obvious, that in general case the initial testing temperature T should be used in the right-hand part of the equation (6.10) instead of the value 293 K.

Hence, the stated above results showed the possibility and expediency of cluster model and fractal analysis usage for polymeric materials friction and frictional wear processes description. Nonlinearity of the temperature in contact zone dependence on testing regimes is defined by polymeric matrix structure fractality. The loosely-packed matrix is the main structural component, responsible to friction and frictional wear processes.

According for the main process character polymers frictional wear can be divided into fatigue and abrasive (microcutting) [14]. The fatigue wear occurs as a result of fatigue fracture and is mainly due to frictional contact discrete character. The abrasive wear contains in cutting process by sharp projections system. The notion "sharp projection" is relative enough, since abrasive wear process depends as well on polymer stiffness and consequently, on its structure state (see section 3.1). Therefore at definite conditions there can be expected transition from one wear mechanism to an other for the same polymeric material without friction tests type change. In the given case it is important that wear abrasive mechanism is one – act contrary to fatigue mechanism [14]. Proceeding from the said above, the authors [15] studied structural changes, defining transition from one wear mechanism to an other on the example of carbon plastics on the basis of phenylone.

For both considered mechanisms the linear wear intensity I is defined by the expression [14]:

$$I = I_1 P^{\alpha}, \qquad (6.11)$$

where I_1 is linear wear intensity at pressure P=1, α is exponent, characterizing polymer nature and surface geometry.

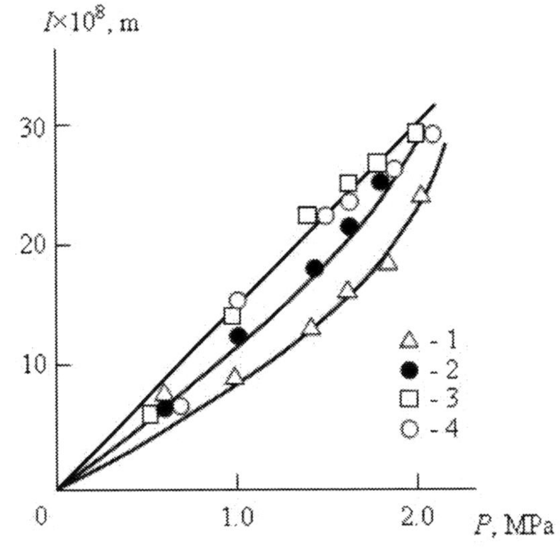

Figure 6.6. The dependence of linear wear intensity I on pressure P for carbon plastics on the basis of phenylone at different sliding rate ϑ: 1.0 (1), 1.5 (2), 2.0 (3) and 2.5 m/s (4) [15].

In usual conditions for fatigue wear $\alpha>1$ and for abrasive one - $\alpha=1$ [14]. Therefore, the exponent α is depending on structure characteristic of polymeric materials wear process and further its correlation possibility with carbon plastics structural characteristics will be studied [15].

In Figure 6.6 the dependence $I(P)$ is adduced for the studied carbon plastics at four sliding rates ϑ within the range of 1.0-2.5 m/s. As it follows from the plots, adduced in this Figure, at small rates (1.0 and 1.5 m/s) the nonlinear dependence $I(P)$ is observed, i.e. according to the equation (6.11) $\alpha>1.0$ and wear mechanism is a fatigue one. At higher ϑ (2.0 and 2.5 m/s) the dependence $I(P)$ is a linear one, $\alpha=1$, that means wear mechanism type change – the transition from the fatigue to the abrasive one [14]. Using generalized characteristic ($p\vartheta$) [5], one can suppose that the fatigue wear mechanism is realized at relatively small values ($p\vartheta$) ($\leq$1.5 MJ/s) and the abrasive one – at higher values. In this case polymeric materials structure change is due to enhancement of contact temperature in friction zone T_c, the value of which is connected with parameter ($p\vartheta$) by the relationship (6.1).

In Figure 6.7 the dependence of the exponent α on cluster relative fraction φ_{cl} is adduced for carbon plastics on the basis of phenylone. As one can see, φ_{cl} increase results to α growth and, accordingly, defines the tendency of transition from the abrasive wear mechanism to the fatigue one. In more general terms the structure change can be described by its fractal dimension d_f variation, which is connected with φ_{cl} according to the equation (1.4). From the data of Figure 6.7 and the equation (1.4) we will obtain, that at $\alpha=0$, i.e. I=const, $d_f\approx2.63$, at $\alpha=1.0$ (abrasive wear mechanism) $d_f=2.47$ and at the greatest possible for phenylone value $\varphi_{cl}=0.74$ $d_f=2.29$ and $\alpha\approx1.5$ (a fatigue wear mechanism).

Let us consider the interconnection of these boundary values of structural characteristic d_f with carbon plastics fracture type. As Balankin showed [9], solids fracture type was defined by their structure, characterized by the value d_f. For $d_f\leq2.50$ brittle fracture is realized, for $2.50<d_f<2.67$ - quasibrittle and for $d_f>2.70$ – ductile one. The comparison of the indicated d_f

values with calculated according to the equation (1.4) boundary values of this dimension for carbon plastics demonstrates that the abrasive wear mechanism corresponds to quasibrittle fracture, the fatigue one up brittle one and α=0 is reached at transition to ductile fracture type, for which *I*≈const, i.e. independent on *P*, that is generally observed for rubbers [14]. Therefore, at any rate for carbon plastics on the basis of phenylone wear mechanism defines by material fracture type [15].

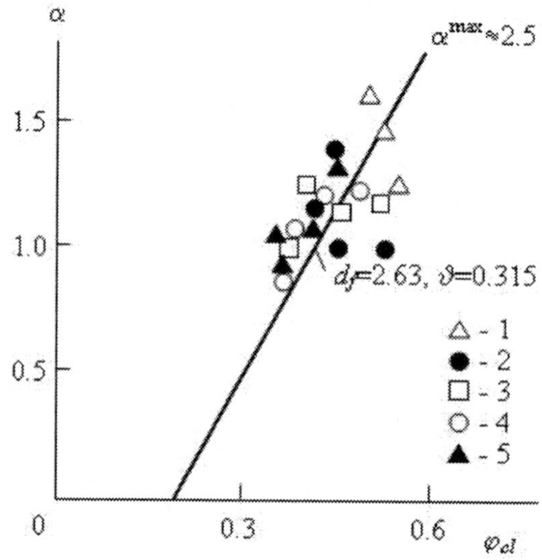

Figure 6.7. The dependence of the exponent α in the equation (6.11) on clusters relative fraction φ_{cl} for carbon plastics on the basis of phenylone at different pressures *P*: 0.6 (1), 1.4 (2), 1.6 (3), 1.8 (4) and 2.0 MPa (5) [15].

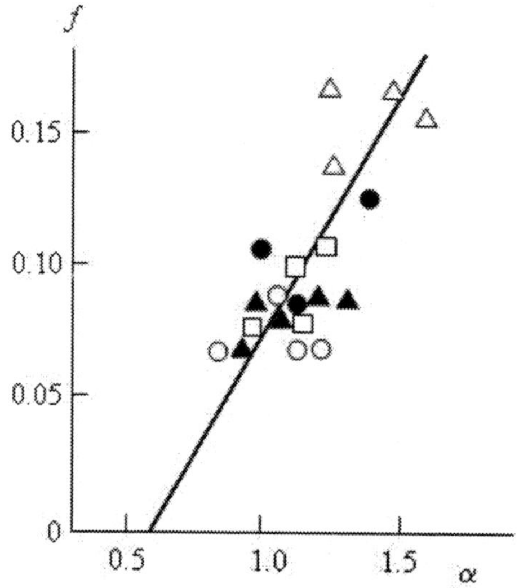

Figure 6.8. The dependence of friction coefficient *f* on the exponent α in the equation (6.11) for carbon plastics on the basis of phenylone. The designations are the same as in Figure 6.7 [15].

The plot of Figure 6.7 allows to determine the limiting values α for carbon plastics on the basis of phenylone. The smallest value α=0 and the greatest value α can be obtained by the linear dependence $\alpha(\varphi_{cl})$ extrapolation to maximum possible value φ_{cl}=0.74 (see the equation (1.10)), that gives $\alpha\approx2.5$. Therefore, for carbon plastics on the basis of phenylone variation α makes up 0-2.5, that corresponds to the known at present data [14].

Polymeric materials wear mechanism and, hence, their fracture type, defines these materials friction coefficient f. In Figure 6.8 the dependence $f(\alpha)$ is adduced, which proves to be linear and demonstrating f growth at α increase. At abrasive wear mechanism (quasibrittle fracture) the value f is relatively small and equal to ~ 0.08 and at the transition to fatigue mechanism the value f increases within the range of 0.08-0.17. The greatest value f for carbon plastics on the basis of phenylone can be estimated by the linear dependence $f(\alpha)$ extrapolation to the maximum value α=2.5, which is equal to ~ 0.31 [15].

Therefore, the stated above results testify that carbon plastics on the basis of phenylone wear mechanism is defined by their structure state in frictional contact zone. This factor defines solids fracture type as well, that allows to assume certain interconnection between wear mechanism and fracture type. In its turn, structure state of carbon plastics, characterized by its fractal (Hausdorff) dimension, is defined by temperature enhancement in the contact zone, depending on friction conditions (sliding rate, specific load). The offered by authors [15] structural treatment allows to ascertain these factors quantitative interconnection.

Polymers reinforcing by the oriented fibers has specific features number in comparison with reinforcing by the particulate mineral fillers [16]. In paper [17] the supposition was stated, that fibers surface fibrous structure results to larger frictional stability in comparison with similar fibers isotropic structure. Besides, the authors [17] demonstrated by IR-spectroscopy methods availability of hydrogen bonds on the division boundary of uglen and vniivlon fibers with polyarylate (PAr), which was used as binding. Such bonds availability assumes PAr macromolecules anisotropy on division boundary, that should be told on composite frictional stability. Such fiber surface main features "imitating" by polymer boundary layers is illustrated well on the example of PAr-terlon system. This fiber has crystalline structure, that results PAr crystallization up to crystallinity degree ~ 0.20 [17], whereas amorphous uglen and vniivlon do not give such effect. Proceeding from the said above, the authors [18, 19] studied surface anisotropy influence of uglen and vniivlon fibers on frictional wear of composites on the basis of PAr filled with the indicated fibers.

In Figure 6.9 the diffraction X-ray curves for PAr, uglen and carbon plastics on the basis of PAr with uglen different volume contents φ_f are adduced. As one can see, the intensity I_X reduction of carbon plastics amorphous halo and its maximum displacement in larger scattering angles 2θ direction is observed at φ_f growth. Such diffraction patterns change indicates on local order domains (clusters) relative fraction φ_{cl} increase in polymeric matrix structure and interchain distance decrease at φ_f growth [20]. Let us note that in the given case clusters generalized definition is accepted: there are structure parts consisting of segments with "freezing" molecular mobility. Besides, composites density ρ measurements indicate on polymeric matrix structure densification at φ_f growth: the experimental value ρ is higher than additive one, this difference grows at φ_f increasing at that [17]. The following relationship was obtained between parameters I_X and φ_{cl} [20]:

$$\varphi_{cl}^{-1} = cI_X. \quad (6.12)$$

where c is constant.

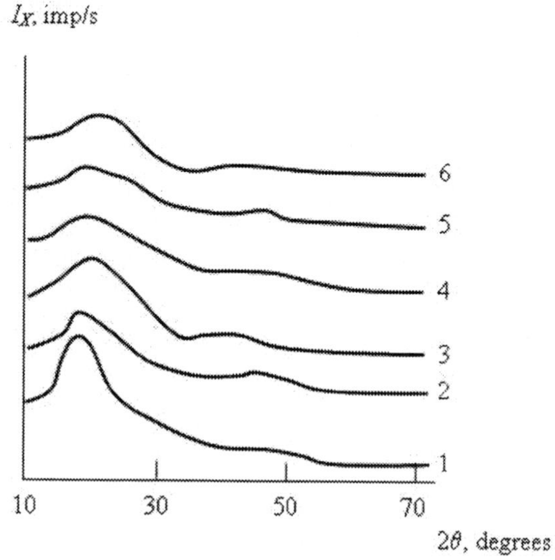

Figure 6.9. The diffraction X-ray curves for polyarylate (1), composites on the basis of polyarylate with uglen volume contents 0.038 (2), 0.114 (3), 0.195 (4), 0.282 (5) and uglen (6) [18].

The value c for the studied composites can be estimated as follows. For the initial PAr the value φ_{cl} can be determined, according to the percolation relationship (1.72). The value c is determined as reciprocal value of product of the obtained in this way φ_{cl} magnitude and experimental value I_X for PAr.

Proceeding from the said above, one can suppose that polymeric matrix structure densification (the estimations according to the equation (6.12) give φ_{cl} enhancement from 0.514 for the initial PAr up to 0.887 for composite PAr-uglen with uglen volume content of 0.282) is defined by macromolecules PAr anisotropy in interfacial layer polymer-filler, which is due to availability of fibers surface structure anisotropy and hydrogen bonds between uglen and PAr [17]. In other words, PAr macromolecules stretching on filler fibers surface is supposed. This stretching extent can be estimated with the aid of the fractal dimension D_{ch} of chain part between its fixation points [8], which can be calculated with the aid of the equation (1.55).

In Figure 6.10 the dependence of the calculated according to the equation (1.55) dimension D_{ch} on uglen and vniivlon volume contents in the considered composites is adduced. As it has been expected, D_{ch} reduction at φ_f growth is observed, assuming PAr chains stretching process on fibers surface (fibrillization). At $\varphi_f \approx 0.25$ asymptotic value $D_{ch} \approx 1.05$, meaning almost complete loss of molecular mobility for PAr chains is reached. Let us note, that for the mentioned composites the smallest value of linear wear intensity I is reached at about $\varphi_f \approx 0.20$-0.22 [17].

Another factor, influencing on polymer fibrillization on the fibers surface, is their total surface area, proportional to φ_f value – the larger this area, the higher concentration of active sites, capable to form hydrogen bonds polymer-filler. Proceeding from the said above, it is expected that D_{ch} decreasing and φ_f enhancement results to I reduction. In Figure 6.11 the dependence of I on ratio D_{ch}/φ_f value is adduced, which approximates well enough by the linear correlation, passing through coordinates origin. Such course of the dependence $I(D_{ch}/\varphi_f)$ shows that frictional wear intensity decreases at the increase of hydrogen bonds number and stretching extent (fibrillization) of polymer chains on anisotropic fibers surface in polymer composites.

Hence, the considered above results showed short fibers surface structure anisotropy influence on interfacial layer structure in polymer composites. In its turn, the indicated structure changes cause these materials frictional wear essential variation. In this aspect hydrogen bonds polymer-fiber availability plays an important role [19].

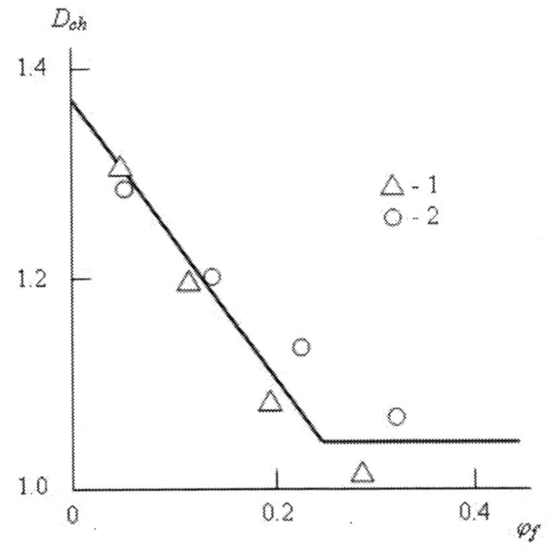

Figure 6.10. The dependence of fractal dimension D_{ch} of chain part between clusters on fibers volume contents φ_f for composites PAr-uglen (1) and PAr-vniivlon (2) [18].

Ultrahighmolecular polyethylene (UHMPE) with molecular weight of ~ 10^6 and higher possesses by a number useful properties. It is notable for high impact toughness, stability to wear, low friction coefficient, self-lubrication, water-resistivity, good dielectric properties, it is harmless and not venomous [21-23]. At special soot addition UHMPE acquires antistatic, current-conducting properties and stability. Aluminium powder or graphite addition increases its thermal conductivity. UHMPE curing by organic peroxides prevents to crystallites formation, reduces density and increases transparency, improves wear-resistence and decreases heat expansion coefficient.

Fillers introduction in UHMPE (for example, chalk, glassy microspheres etc.) improves its thermophysical properties, raises hardness, stiffness and strength at mechanical loading [24]. In paper [25] it was shown that 40 mass % high-modulus carbon fibers introduction in UHMPE allowed to reduce frictional wear intensity in about 40 times – from 5 mcm/km for initial UHMPE up to 0.12 mcm/km for carbon plastic on the basis of UHMPE. The authors [26] studied the physical causes of this important effect both practically and theoretically.

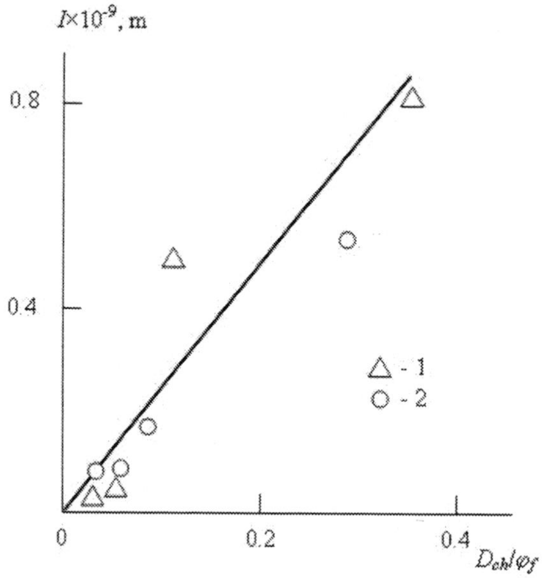

Figure 6.11. The dependence of linear wear intensity I on ratio (D_{ch}/φ_f) value for composites PAr-uglen (1) and PAr-vniivlon (2) [18].

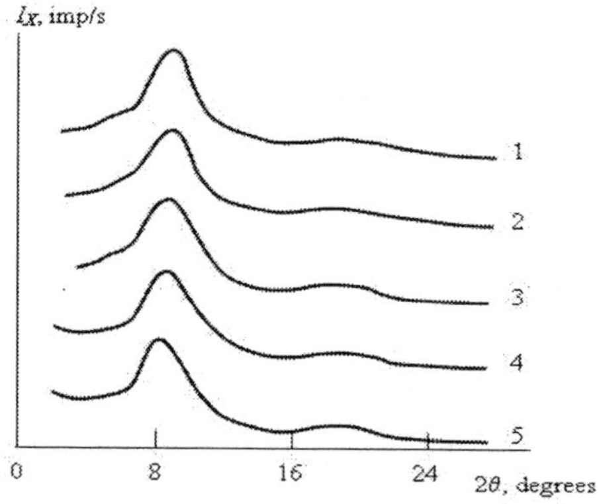

Figure 6.12. X-ray diffractograms of UHMPE (1) and carbon plastics on its basis with carbon fiber contents 5 (2), 10 (3), 15 (4) and 20 (5) mass. % [26].

The structure of carbon plastics on the basis of UHMPE studies by traditional methods (wide-angle X-raying, IR-spectroscopy) did not find essential changes in comparison with structure of initial matrix polymer. As it follows from the data of Figure 6.12 for UHMPE and carbon plastics the identical X-ray diffractograms were obtained. The results of the studies by IR-spectroscopy method show chemical interaction absence between carbon fiber and UHMPE, since the bonds position in region with maximum at 2960-2880 cm^{-1}, connected with valent oscillations of UHMPE methylene groups C-H bonds, is preserved independently on components quantitative relation (Figure 6.13).

Nevertheless, despite the adduced above results of X-ray diffractometry and IR-spectroscopy, strong enough changes of carbon plastics properties are observed at carbon fiber contents increasing (table 6.1). The values of two parameters change particularly significantly: microhardness *HB* (about twice) and linear frictional wear intensity *I* (in about 40 times). The indicated in table 6.1 carbon plastics properties change at structure integral characteristics approximate equality assumes their morphology essential changes, which are due to filler introduction. It's possible to trace the assumed morphological changes most simply and visually by microhardness *HB* variation using [27, 28]. For semicrystalline polymers the value *HB* can be written as the sum [27]:

$$HB = K \cdot HB^{cr} + (1-K)HB^{am}, \tag{6.13}$$

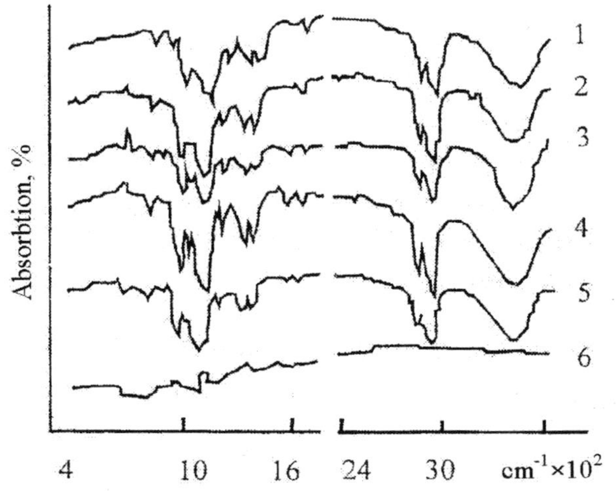

Figure 6.13. IR-spectra of absorbtion for UHMPE (1), carbon plastics on its basis with carbon fiber contents 5 (2), 10 (3), 15 (4) and 20 (5) mass. % and carbon fiber "Ural 15" (6) [26].

where K is crystallinity degree, HB^{cr} and HB^{am} are crystalline and amorphous phases microhardness, accordingly.

Since for UHMPE amorphous phase at testing temperature (T=293 K) is devitrificated, then for it HB^{am}=0. This rule for UHMPE is confirmed experimentally [29] and then the equation (6.13) is simplified up to [26]:

$$HB = K \cdot HB^{cr}. \tag{6.14}$$

For UHMPE the value K=0.42 was obtained by wide-angle X-raying method [29]. The calculation by polymer density gives close value of crystallinity degree (K=0.48) [30]. Then the value HB^{cr}, calculated according to the equation (6.14), is equal to 224 MPa.

The microhardness and polymer morphology interconnection can be expessed as follows [27]:

$$HB = \frac{HB^{cr}}{1 + \left(b / \langle l_{cr} \rangle\right)}, \tag{6.15}$$

Table 6.1. The carbon fiber contents influence on properties of carbon plastics on the basis of ultrahighmolecular polyethylene [25]

Carbon fibers contents, mass. %	*HB*, MPa	σ_Y^{comp}, MPa	E_c, MPa	*I*, mcm/km	*f*
0	94	64.0	1370	5.0	0.50
5	110	71.0	1600	3.2	0.42
10	120	72.5	1700	2.0	0.35
15	138	75.5	1800	1.2	0.33
20	175	79.5	1800	0.25	0.28
30	192	87.0	2010	0.15	0.29
40	186	103.0	1460	0.12	0.34

Notes: *HB* is Brinell microhardness, σ_Y^{comp} is yield stress at compression, E_c is elasticity modulus, *I* is linear frictional wear intensity, *f* is friction coefficient.

where *b* is constant, equal to 20 nm, $\langle l_{cr}\rangle$ - the average thickness of crystalline lamellas.

The equation (6.15) allows $\langle l_{cr}\rangle$ calculation according to the known *HB* values. These estimations showed $\langle l_{cr}\rangle$ increase from 14.5 nm for the initial UHMPE that corresponds to the experimental data [31], up to 120 nm for carbon plastics, containing 30 mass % carbon fibers. Such variation of carbon plastics morphology is law-governed. As it is known [32], macromolecules of matrix polymer on carbon fibers smooth surface change their conformation and stretch, forming densely-packed interfacial layer. Such layer crystallization results to strongly stretched lamellas formation, that defines $\langle l_{cr}\rangle$ growth at fibers contents increase (crystallization mechanism change, see section 1.4). As it was noted in paper [17], fibrous (consisting of stretched macromolecules) polymers structure raised frictional wear-resistance in comparison with isotropic structure of the same polymer. Therefore one should expect that $\langle l_{cr}\rangle$ increase results to linear frictional wear intensity *I* reduction.

The second factor, influencing on the value *I*, is bond polymer-filler availability. Chemical bonds between UHMPE matrix and carbon fibers absence, which IR spectroscopy data (Figure 6.13) show, does not reject intermolecular van-der-waals bonds formation possibility on the indicated division boundary. It is obvious, such bonds number will be proportional to fibers surface total area, which, in its turn, is proportional to $\varphi_f^{2/3}$, where φ_f is filler volume contents. An indicated bonds number increase will raise polymeric matrix – fiber interaction extent and reduce the value *I*. Proceeding from the said above, it is expected that *I* value will be smaller at larger product ($\langle l_{cr}\rangle\varphi_f^{2/3}$). Therefore in Figure 6.14 the dependence of linear frictional wear intensity *I* on reciprocal value of the indicated parameter ($\langle l_{cr}\rangle\varphi_f^{2/3}$) for carbon plastics on the basis of UHMPE is adduced. As one can see, the plot of Figure 6.14 shows *I* linear reduction at product ($\langle l_{cr}\rangle\varphi_f^{2/3}$) increase, and approximating this dependence straight line passes through coordinates origin. It is significant, that product ($\langle l_{cr}\rangle$

$\varphi_f^{2/3}$) increase in about 20 times gives the same reduction of frictional wear intensity for the considered carbon plastics.

Hence, the adduced above data have demonstrated that filler fibers introduction in semicrystalline polymer, without changing its integral structural characteristic (crystallinity degree) can essentially change the crystalline phase morphology. Besides, the chemical bonds polymer-filler absence preserves the possibility of intermolecular van-der-waals bonds formation on the indicated boundary. The both mentioned factors essentially influence on frictional wear value of carbon plastics.

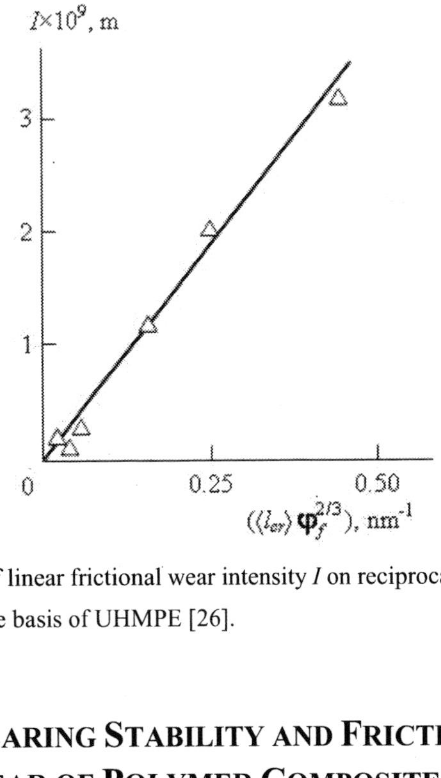

Figure 6.14. The dependence of linear frictional wear intensity I on reciprocal value of product ($\langle l_{cr}\rangle$ $\varphi_f^{2/3}$) for carbon plastics on the basis of UHMPE [26].

6.2. Shearing Stability and Frictional Wear of Polymer Composites

The stated above structural models are not only the ones, used for polymer composites friction and wear processes description [33, 34]. At present it is known [9], that solid deformation character depends on nondeformed material shearing stability and its deformation conditions. For shearing stability level estimation the parameter R_{cr}, similar to Reynolds number at turbulent flow, is determined as follows [9]:

$$R_{cr} = \frac{\sigma_{\lim}}{2\tau_{\lim}}, \tag{6.16}$$

where $\sigma_{\lim}$ and $\tau_{\lim}$ are limiting normal and shear stresses, accordingly.

It was shown [9], that in case of shear-stable materials $R_{cr}<1/\sqrt{2}$ and in case of shear-instable ones – $R_{cr}>1/\sqrt{2}$. For independent R_{cr} determination the fact, that deformation resistance in time initial moment can be used, when plastic deformation has still laminar character, is defined by yield stress σ_Y and in turbulent regime, realized at large strains (that is true to full extend and for polymers [35]) – by microhardness *HB*. Hence the relationship between σ_Y and *HB* follows, having the form [9]:

$$\frac{HB}{\sigma_Y}=1+\frac{2}{R_{cr}}. \quad (6.17)$$

Combination of the equations (6.16) and (6.17) allows to calculate the value τ_{lim} according to experimentally determined values *HB* and σ_Y. In Figure 6.15 the dependence of τ_{lim} on lamellas average thickness $\langle l_{cr}\rangle$, determined according to the equation (6.15), for carbon plastics on the basis of UHMPE is adduced, where the boundary between shear-stable and shear-instable polymeric matrix is shown by a horizontal shaded line. As it follows from the data of this Figure, $\langle l_{cr}\rangle$ increase results to τ_{lim} fast growth and at $\langle l_{cr}\rangle$>40 nm the transition to shear-stable polymeric matrix is observed. It is important to note, that approximately the same boundary divides fatigue and abrasive mechanisms of frictional wear [34].

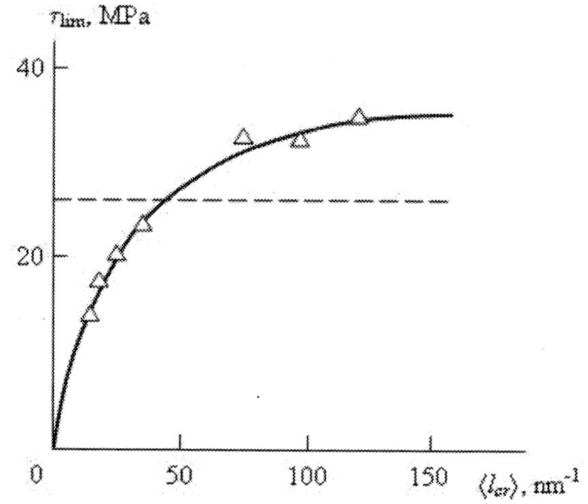

Figure 6.15. The dependence of limiting shear stress τ_{lim} on thickness $\langle l_{cr}\rangle$ for carbon plastics on the basis of UHMPE. The horizontal shaded line indicates boundary between shear-stable and shear-instable polymeric matrix [34].

In Figure 6.16 the dependence of linear frictional wear intensity *I* on τ_{lim} is shown, from which fast *I* reduction at τ_{lim} growth follows. In Figure 6.16 the boundary between shear-instable and shear-stable polymeric matrix is shown by a vertical shaded line. One can see, that this line divides two regions of the dependence $I(\tau_{\text{lim}})$: in the first case fast *I* reduction at τ_{lim} growth is observed and in the second one practically asymptotic (slow) *I* change is obtained.

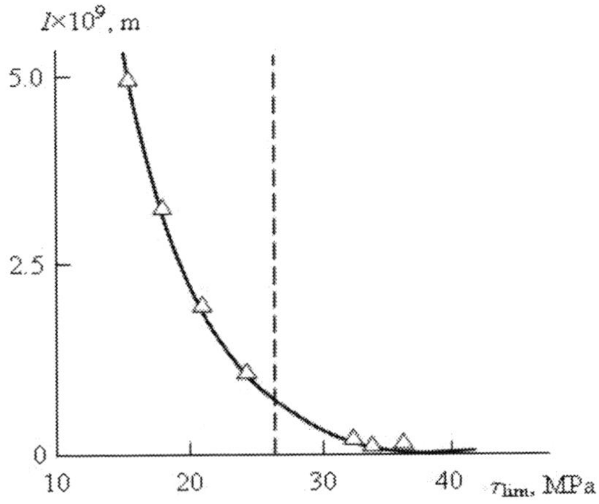

Figure 6.16. The dependence of linear frictional wear intensity I on limiting shear stress $\tau_{\rm lim}$ for carbon plastics on the basis of UHMPE. The vertical shaded line indicates boundary between shear-instable and shear-stable polymeric matrix [34].

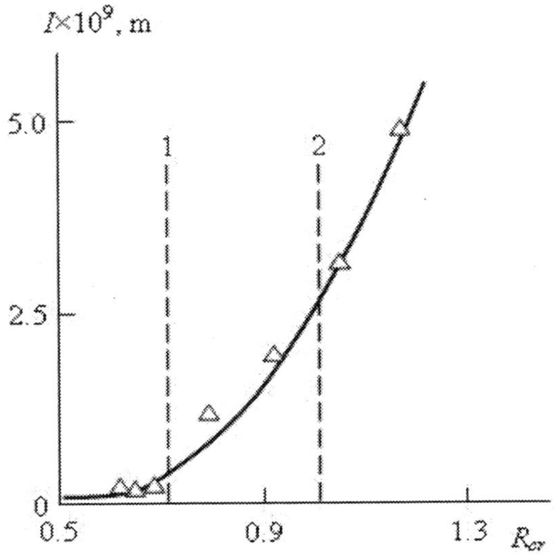

Figure 6.17. The dependence of linear frictional wear intensity I on parameter R_{cr}. The vertical shaded lines indicate the transitions from brittle to brittle-ductile (1) and from brittle-ductile to ductile (2) fracture types [33].

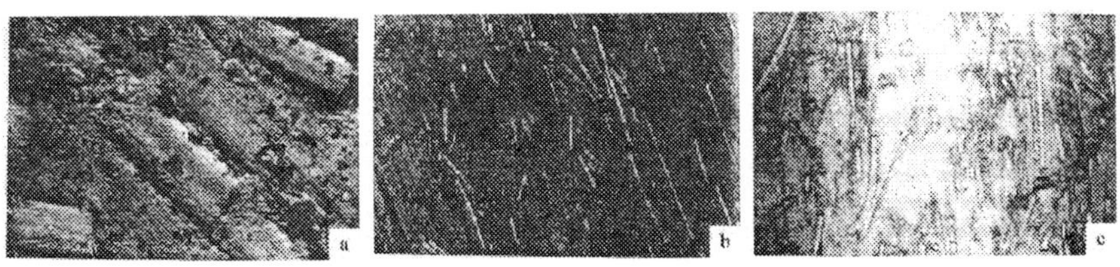

Figure 6.18. The optical micrographs of friction surface of UHMPE (a, 3000×) and carbon plastics, containing 10 (b, 500×) and 20 (c, 500×) mass. % carbon fibers [33].

The materials fracture different types can be identified with the aid of the parameter R_{cr} as follows: for $R_{cr}<1/\sqrt{2}$ fracture will be brittle, for $1/\sqrt{2}<R_{cr}<1$ – brittle-ductile and for $R_{cr}>1$ – ductile one. In Figure 6.17 the dependence $I(R_{cr})$ is shown and the cited above fracture different types boundaries [9] are indicated by vertical shaded lines. From the data of Figure 6.17 it follows that the greatest and reducing rapidly at R_{cr} decreasing carbon plastics frictional wear is observed at ductile type of fracture and practically constant – at brittle one.

The adduced above results are confirmed by the data of optical microscopy [25, 33]. A transfer and spreading of polymer on contrabody are observed; explosion zones appear on UHMPE samples friction surface (ductile fracture). The introduction of carbon fibers in to polymer contributes to polymer transfer cessation. Friction of carbon plastics containing 5-10 mass % filler, is accompanied by fibers orientation along friction direction (brittle-ductile fracture). At carbon fibers contents increase within the range of 20-40 mass. % the tendency to preserve by fiber its initial chaotic distribution, fixed at preparation, appears (brittle fracture, Figure 6.18).

The obtained above correlations allow to predict the value I for the considered carbon plastics as the function of structural characteristics, for example $\langle l_{cr}\rangle$, one common equation being used for both fatigue and abrasive wear mechanisms. For the studied carbon plastics the authors [34] obtained the following relationship between I and $\langle l_{cr}\rangle$:

$$I = 10^{-9}\left(0.0308\langle l_{cr}\rangle\right)^{-2}, \qquad (6.18)$$

where I is given in m and $\langle l_{cr}\rangle$ - in nm.

In Figure 6.19 the comparison of experimentally determined I values and calculated according to the equation (6.18) I^T magnitudes of linear frictional wear intensity for the considered carbon plastics on the basis of UHMPE is adduced. As one can see, the good correspondence is obtained between experiment and theory (the average discrepancy between I and I^T does not exceed 15 % that is comparable with the experimental error at this parameter determination). Let us note, that comparison of Figures 6.14 and 6.19 plots assumes parameters $\langle l_{cr}\rangle$ and $\varphi_f^{2/3}$ interconnection or; more precisely, their proportionality.

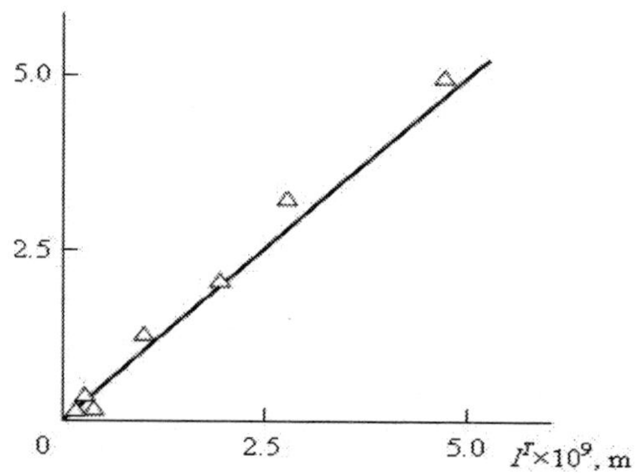

Figure 6.19. The relation of experimentally determined I and calculated according to the equation (6.18) I^T values of linear frictional wear intensity for carbon plastics on the basis of UHMPE [34].

Hence, the results stated above showed that the main structural characteristic, defining frictional wear value, was polymeric matrix shearing stability. This criterion is true for both fatigue and abrasive mechanisms of frictional wear. The introduction of short fibers with relatively smooth surface into the polymer creates densely-packed interfacial regions, which are deformed by shearing mechanism, thus reducing shearing stability of carbon plastics binding and decreasing their frictional wear [33, 34].

The frictional wear process of polymeric materials is of great importance from practical point of view and is very complex for the research from theoretical point of view [14]. This complexity results to the appearance of theoretical conceptions number, describing frictional wear process of polymeric materials within the frameworks of two wear mechanisms – fatigue and abrasive ones, criterion of which division is the equation (6.11). For the fatigue wear description the following formula can be used [14]:

$$I \approx \frac{P}{n_d HB}, \qquad (6.19)$$

where P is pressure, n_d is limiting number of surface layers deformation cycles, HB is microhardness.

At the same time for abrasive wear the following equation was offered [14]:

$$I \approx k_a \frac{P}{HB}, \qquad (6.20)$$

where k_a is coefficient.

From the comparison of the equations (6.19) and (6.20) their identity follows at $k_a = n_d^{-1}$. Proceeding from the said above, one can expect that some structural characteristic exists, with the aid of which frictional wear intensity can be described independently from its mechanism. One of the possible variants is the parameter R_{cr} using, which was considered in paper [36] on the example of carbon plastics on the basis of phenylone.

Besides the described above calculation R_{cr} method with the aid of the equation (6.17), another variant of this parameter estimation gives the equation [9]:

$$R_{cr} = \frac{\nu B'}{4(1+\nu)^{1/2}}, \qquad (6.21)$$

where B is equal to [9]:

$$B' = \left(\frac{\partial B}{\partial P}\right)_T = 2(m+n+1). \qquad (6.22)$$

In the equation (6.22) the expression in brackets in its left-hand part represents an isothermal derivative of bulk modulus B by pressure P and m and n are exponents in Lennard-Jones equation, which in its turn depend on material structural state. For polymers this dependence has the form [37]:

$$mn = \frac{6(1+\nu)}{1-2\nu}, \tag{6.23}$$

where ν is Poisson's ratio.

For metals correctness of the following equation was shown [9]:

$$mn = \nu^{-1}. \tag{6.24}$$

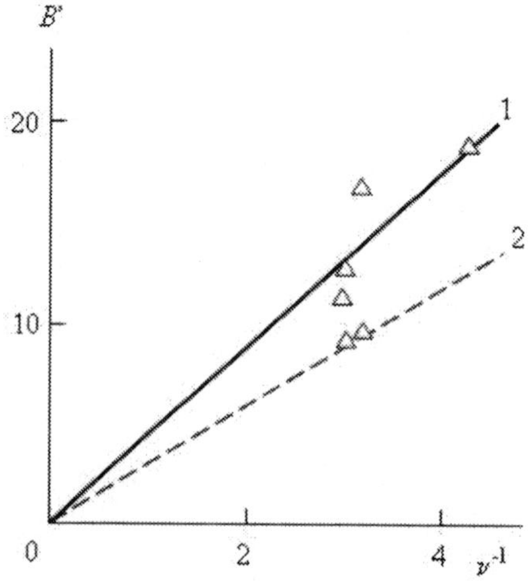

Figure 6.20. The dependence of parameter B' on Poisson's ratio ν reciprocal value for carbon plastics on the basis of UHMPE (1) and metals (2) [36].

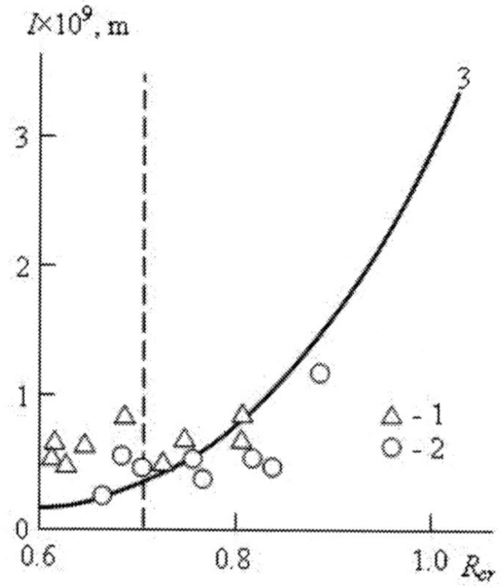

Figure 6.21. The dependence of linear frictional wear intensity I on parameter R_{cr} value for carbon plastics on the basis of phenylone, prepared with magnetic (1) and mechanical (2) separation application and also on the basis of UHMPE (3). The vertical shaded line indicates boundary between shear-instable and shear-stable polymeric matrices [36].

The distinction of the equations (6.23) and (6.24) is defined by the interatomic bonds type distinction in the indicated materials: van-der-waals and metallic ones, accordingly. Using the data for carbon plastics on the basis of UHMPE [25] and the equations (6.17) and (6.21) and also determining the value ν according to the equation (1.2), the authors [36] plotted the dependence $B'(\nu^{-1})$ for these composites, which was shown in Figure 6.20. As one can see, the indicated dependence corresponds to the equation (6.24). Besides, in the same Figure the dependence $B'(\nu^{-1})$ for metals (W, Mo, Ta, V, Nb), plotted according to the data of work [9], is shown by a shaded line, which is also linear, but gives the value B' somewhat smaller, than a similar dependence for polymers. The specific feature of the dependence $B'(\nu^{-1})$ for carbon plastics on the basis of UHMPE is different values B' for brittle and ductile types of fracture: for the ductile one (ν>0.25, ν^{-1}<4) the value $B'\approx19$. Further just these values B' were used for the parameter R_{cr} calculation in case of carbon plastics on the basis of phenylone [36].

In Figure 6.21 the dependence of linear frictional wear intensity I on R_{cr} value for carbon plastics on the basis of phenylone (points) and UHMPE (solid line) is adduced. As one can see, the data for these two composites coincide both qualitatively and quantitatively, despite the large distinction of matrix polymers structure and properties. Besides, carbon plastics on the basis of UHMPE contain carbon fiber within the range of 5-40 mass. % and for carbon plastics on the basis of phenylone carbon fiber contents is constant and equal to 15 mass. %. In Figure 6.21 the boundary between shear-instable and shear-stable polymeric matrices is indicated by the vertical shaded line: as it was noted above, for the first $R_{cr}>1/\sqrt{2}$, for the second - $R_{cr}<1/\sqrt{2}$ [9]. As one can see, for shear-stable matrixes the values I are small (< 0.8×10^{-9} m) and approximately constant, whereas for shear-instable matrixes I rapid growth is observed at R_{cr} increasing.

Hence, the stated above results showed that the criterion R_{cr} can be a general parameter, characterizing polymeric materials frictional wear independently on this wear mechanism. This parameter gives polymeric matrix shear stability measure independently on its nature, filler contents and so on. To obtain wear-resistant composites realization of shear-instable polymeric matrix, i.e. having the value $R_{cr}<1/\sqrt{2}$ is required [9].

Therefore, the studies of carbon plastics on the basis of UHMPE and phenylone showed linear frictional wear intensity I reduction at the parameter R_{cr} decreasing [33, 36]. The authors [38] considered this postulate universality and its application possibility for the value I prediction in case of carbon plastics on the basis of phenylone with variable testing conditions – pressure P and sliding rate ϑ [2]. It was shown earlier [5], that temperature T_c in the friction zone can be calculated according to the equations (6.1) and (6.10). The combination of the mentioned equations allows to obtain the relationship for determination of carbon plastics structure fractal dimension d_f as a function of $(p\vartheta)$ [38]:

$$d_f = d - \frac{72}{113(p\vartheta)^{1/2}}, \quad (6.25)$$

where d is dimension of Euclidean space, in which fractal is considered (it is obvious, that in the given case d=3).

It is also obvious, that d_f increase at $(p\vartheta)$ growth is due to T_c enhancement. In its turn, Poisson's ratio ν value can be determined according to the known d_f values with the aid of the equation (1.1) and the parameter R_{cr} can be further calculated according to the equations (6.21)-(6.23), where as ν minimum for carbon plastics on the basis of phenylone Poisson's ratio value (ν=0.146 [39]) was used.

In Figure 6.22 the dependence of linear frictional wear intensity I on the parameter R_{cr} for carbon plastics on the basis of phenylone at variation P=1.0-2.0 MPa and ϑ=1.0-2.5 m/s, i.e. at variation $(p\vartheta)$=1.0-5.0 MPa·m/s is adduced. As one can see, I growth at R_{cr} increase is observed again. In the same Figure the dependence $I(R_{cr})$ for carbon plastics on the basis of UHMPE is shown by the solid curve, which quantitatively corresponds well enough to similar data for carbon plastics on the basis of phenylone and in essence is a lower boundary for the latter. This correspondence can be improved by the value B' variation in the equation (6.21): $(p\vartheta)$ increase results to ν growth and, according to the equation (6.23), to product mn enhancement and, accordingly, the sum $(m+n+1)$. Besides, as it noted has been above, the materials fracture different types can be identified with the aid of parameter R_{cr} as follows: for $R_{cr}<1/\sqrt{2}$ fracture will be brittle, for $1/\sqrt{2}<R_{cr}<1.0$ – brittle-ductile and for $R_{cr}>1.0$ – ductile one [9]. In Figure 6.22 these boundaries are shown by the vertical shaded lines, from which it follows that fracture of carbon plastics on the basis of phenylone in the studied range of $(p\vartheta)$ has brittle-ductile character.

The data of Figure 6.22 allow to obtain the approximated relationship between I and R_{cr} in the form [38]:

$$I \approx 6.3\left(R_{cr}^2 - 0.36\right), \text{ nm.} \tag{6.26}$$

The indicated equation together with the relationships (1.1), (6.21) and (6.25) can be used for the value I prediction as a function of $(p\vartheta)$. In Figure 6.23 the comparison of experimental I and calculated by the indicated way I^T linear frictional wear intensity values for carbon plastics on the basis of phenylone and UHMPE is adduced. As one can see, the good enough correspondence of theory and experiment is obtained – the average discrepancy of I and I^T makes up 22 % that satisfies to the requirements of similar characteristics prediction. It is significant that the data for carbon plastics on the basis of UHMPE (which by their characteristics differs essentially from phenylone) with variable carbon fiber contents and at $(p\vartheta)$=const=0.9 MPa·m/s give the correspondence to theory with about the same average discrepancy. This speaks about the universality of the chosen parameter R_{cr}.

Hence, the authors [38] offered prediction methodics of linear frictional wear intensity as a function of testing conditions (pressure and sliding rate). This methodics is based on the structural characteristics usage and even in its most simple form gives a good correspondence to experiment. This means, that polymeric matrix stability to shearing deformation is a universal characteristic for polymeric materials frictional wear process description.

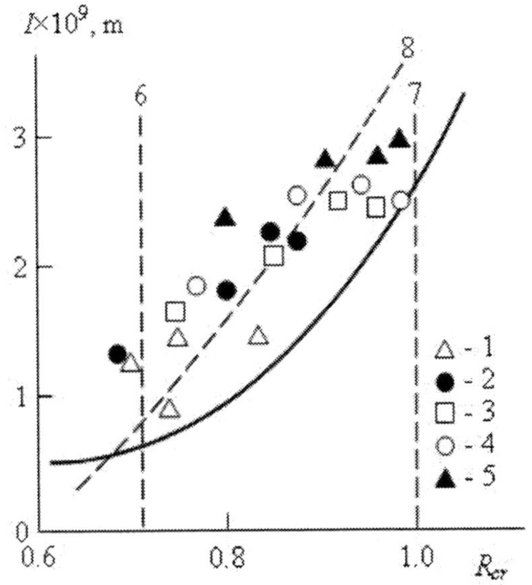

Figure 6.22. The dependence of linear frictional wear intensity I on the parameter R_{cr} for carbon plastics on the basis of phenylone at pressure P: 1.0 (1), 1.4 (2), 1.6 (3), 1.8 (4) and 2.0 MPa (5). The solid line is the data for carbon plastics on the basis of UHMPE. The vertical shaded lines are the indicated transition from brittle to brittle-ductile (6) and from brittle-ductile to ductile (7) fracture types [38].

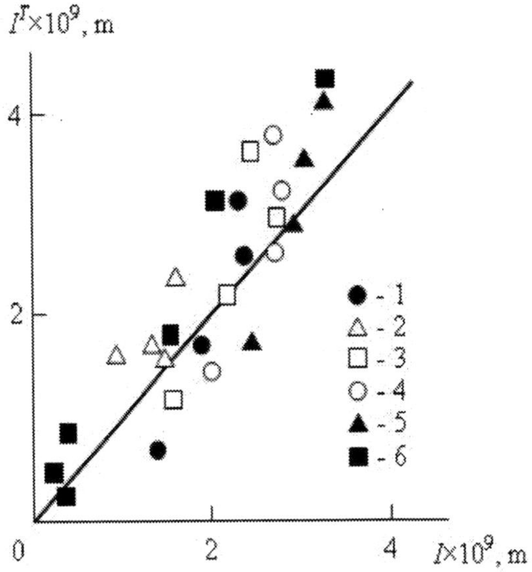

Figure 6.23. The comparison of experimental I and calculated according to the equation (6.26) I^T linear frictional wear intensity values for carbon plastics on the basis of phenylone at pressure P: 1.0 (1), 1.4 (2), 1.6 (3), 1.8 (4) and 2.0 MPa (5) and carbon plastics on the basis of UHMPE (6). The straight line gives relation 1:1 [38].

6.3. THE ADAPTIVITY IN FRICTION PROCESSES OF POLYMER COMPOSITES

As it was shown in paper [5], structure changes of polymers, which are thermodynamically nonequilibrium solids, in friction and wear processes can be influenced essentially on these processes characteristics. The synergetics laws postulate that structures self-organization effects, resulting to their adaptation, can be arisen only in nonequilibrium conditions [40]. Proceeding from these main principles, the authors [41] used synergetics methods for friction and wear processes description for carbon plastics on the basis of phenylone.

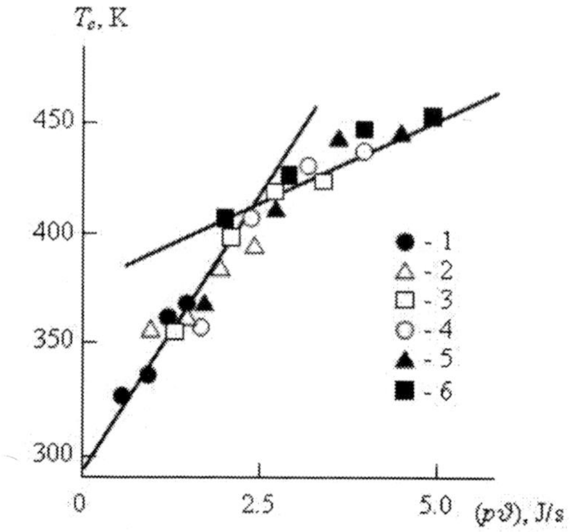

Figure 6.24. The dependence of temperature in contact zone T_c on the value $(p\vartheta)$ for carbon plastics on the basis of phenylone at specific loads p: 0.6 (1), 1.0 (2), 1.4 (3), 1.6 (4), 1.8 (5) and 2.0 MPa (6) [41].

In Figure 6.24 the dependence of temperature in contact zone T_c on product $(p\vartheta)$, where p is normal load, ϑ is sliding rate, for carbon plastics on the basis of phenylone is shown, similar to the adduced one in Figure 6.1. However, unlike Figure 6.1, now the dependence $T_c(p\vartheta)$ is approximated by two linear sections. Therefore, at certain critical temperature T_c^{cr} the dependence $T_c(p\vartheta)$ changes spontaneously, i.e. the temperature T_c^{cr} corresponds to the bifurcation point, at which reaching the old structure becomes instable and in adaptation process the new structure, more stable to temperature action, is self-organized. The universal adaptation algorithm, described by the equation (1.17), is realized at the mentioned transition.

Using as critical values of governing parameter the boundary magnitudes of temperature T_c, the authors [41] obtained the following values of friction process synergetic characteristics: for the range of T_c=327-413 K A_m=0.711, Δ_i=0.255 and m=1 and for T_c=413-452 K A_m=0.925, Δ_i=0.255 and m=16, determinable by gold proportion law [40]. The following conclusion can be made from these results. The transition to the more adapted structure at the expence of reshaping number m increasing at the structure constant stability Δ_i. As the electron microscopy data have shown, within the range of T_c=413-452 K filler fibers parallel drawing up is observed, whereas at T_c<413 K the fibers distribution has chaotic character. It is obvious, that the indicated drawing up of fibers is that structure change, which results to reshaping number *m* enhancement. In addition linear frictional wear increase at T_c growth is sharply decelerated, i.e. the system (structure) is adapted to temperature T_c growth [41].

REFERENCES

[1] Burya A.I., Levi A.G., Bedin A.S., Levit R.M., Raikin V.G. *Trenie i iznos,* 1984, v. 5, № 5, p. 932-935.

[2] Burya A.I., Molchanov B.I. *Trenie i iznos,* 1992, v. 13, № 5, p. 900-904.

[3] Kozlov G.V., Novikov V.U. *Uspekhi Fizicheskikh Nauk,* 2001, v. 171, № 7, p. 717-764.

[4] Kozlov G.V., Zaikov G.E. Structure of the Polymer Amorphous State. Utrecht-Boston, *Brill Academic Piblishers,* 2004, 465 p.

[5] Burya A.I., Kozlov G.V. *Trenie i iznos,* 2003, v. 24, № 3, p. 279-283.

[6] Meilanov R.P., Sveshnikova D.A., Shabanov O.M. Izvestiya VUZov, Severo-Kavkazsk. *region, estestv. nauki,* 2001, № 1, p. 63-66.

[7] Novikov V.U., Kozlov G.V. *Mekhanika Kompozitnykh Materialov,* 1999, v. 35, № 3, p. 269-290.

[8] Kozlov G.V., Novikov V.U. *Synergetics and Fractal Analysis of Cross-Linked Polymers*. Moscow, Klassika, 1998, 112 p.

[9] Balankin A.S. *Synergetics of Deformable Body.* Moscow, Publishers Ministry Defence SSSR, 1991, 404 p.

[10] Adams G.W., Farris R.J. *J. Polymer Sci.: Part B: Polymer Phys.*, 1988, v. 26, № 4, p. 433-445.

[11] Kozlov G.V., Zaikov G.E. The structural Stabilization of Polymers: Fractal Models. Utrecht-Boston, *Brill Academic Piblishers,* 2006, 345 p.

[12] Nigmatullin R.R. *Teoreticheskaya i Matematicheskaya Fizika,* 1992, v. 90, № 3, p. 354-367.

[13] Kozlov G.V., Sanditov D.S., Ovcharenko E.N. Proceedings of International Interdisciplinary Seminar "Fractals and Applied Synergetics, FiPS-01". Moscow, *Publishers MSOU,* 2001, p. 81-83.

[14] Bartenev G.M., Lavrent'ev V.V. Friction and Wear of Polymers. Leningrad, *Khimiya,* 1972, 240 p.

[15] Burya A.I., Kozlov G.V. *Trenie i iznos,* 2005, v. 26, № 3, p. 321-324.

[16] Lipatov Yu.S. Interfacial Phenomena in Polymers. Kiev, *Naukova Dumka,* 1980, 260 p.

[17] Burya A.I., Chigvintseva O.P., Suchilinz-Sokolenko S.P. Polyarylates. Synthesis, Properties, Composite Materials. Dnepropetrovsk, *Nauka i Obrazovanie,* 2001, 152 p.

[18] Kozlov G.V., Burya A.I., Zaikov G.E. *J Appl. Polymer Sci.,* 2006, v. 100, № 4, p. 2821-2823.

[19] Kozlov G.V., Burya A.I., Zaikov G.E. In book: Chemical Reactions in Condensed Phase. Quantitative Level. Ed. Zaikov G., Zaikov V., Mikitaev A. New York, *Nova Science Publishers, Inc.*, 2006, p. 201-206.

[20] Kozlov G.V., Kuznetsov E.N., Beloshenko V.A., Lipatov Yu.S. *Doklady NAN Ukraine,* 1995, № 11, p. 102-104.
[21] Properties and Application of Ultrahighmolecular Polyethylene. Heideekesu, *Haikan to Somu,* 1978, v. 18, № 7, p. 38-47.
[22] Phateja S.K., Reike J.K., Andrews E.H. *J. Mater. Sci.,* 1979, v. 14, № 9, p. 2103-2109.
[23] Dowson D., Challen J.M., Holmes K., Atkinson J.P. Mater. Proc. 3-rd Tribological Sump. *"Wear Non-Metal. Mater."* Leeds-Lyon, London, 1978, p. 99-102.
[24] Braun G., Trauben J. *Kunststoffe,* 1979, H. 69, № 8, p. 434-439.
[25] Burya A.I. NORDTRIB'98: *Proc. Of the 8 International Conf. on Tribology, v. 1.* Ebeltoft, Denmark, 7-10 June 1998, p. 217-220.
[26] Kozlov G.V., Burya A.I., Zaikov G.E. J Appl. Polymer Sci., 2004, v. 93, № 5, p. 2352-2355.
[27] Balta-Calleja F.J., Kilian H.-G. *Colloid Polymer Sci.,* 1988, v. 266, № 1, p. 29-34.
[28] Balta-Calleja F.J., Santa Cruz C., Bayer R.K., Kilian H.-G. *Colloid Polymer Sci.,* 1990, v. 268, № 5, p. 440-446.
[29] Aloev V.Z., Kozlov G.V. Physics of Orientational Phenomena in Polymeric Materials. *Nal'chik, Polygraphservice and T.,* 2002, 288 p.
[30] Kozlov G.V., Beloshenko V.A., Varyukhin V.N., Novikov V.U. *Zhurnal Fizicheskikh Issledovamiy,* 1997, v. 1, № 2, p. 204-207.
[31] Kozlov G.V., Aloev V.Z., Novikov V.U., Beloshenko V.A., Zaikov G.E. Plast. *Massy,* 2001, № 3, p. 21-23.
[32] Pfeifer P. In book: *Fractals in Physics.* Ed. Pietronero L., Tosatti E. Amsterdam, Oxford, New York, Tokyo, North-Holland, 1986, p. 72-81.
[33] Burya A.I., Kozlov G.V. Proceedings of WTC-2005 *World Tribology Congress III,* Sept. 12-16, 2005, Washington, D.C., USA, p. 1-2.
[34] Burya A.I., Kozlov G.V., Kholodilov O.V. *Trenie i iznos,* 2005, v. 26, № 4, p. 407-411.
[35] Gazaev M.A., Kozlov G.V., Mil'man L.D., Mikitaev A.K. *Fizika i Tekhnika Vysokikh Davleniy,* 1996, v. 6, № 1, p. 76-81.
[36] Burya A.I., Kozlov G.V. *Problemy Tribologii,* 2005, № 1, p. 139-142.
[37] Sanditov D.S., Bartenev G.M. Physical Properties of Non-Ordered Structures. Novosibirsk, *Nauka,* 1982, 259 p.
[38] Burya A.I., Kozlov G.V., Rula I.V. *Trenie i iznos,* 2005, v. 26, № 2, p. 187-190.
[39] Burya A.I., Kozlov G.V. *Voprosy Khimii i Khimicheskoy Tekhnologii,* 2005, № 3, p. 106-112.
[40] Ivanova V.S. Synergetics. Strength and Fracture of Metallic Materials. Moscow, *Nauka,* 1992, 155 p.
[41] Burya A.I., Kozlov G.V., Novikov V.U. Proceedings of International Interdisciplinary Symposium "Fractals and Applied Synergetics, FiPS-03". Moscow, *Publishers MSOU,* 2003, p. 38-39.

INDEX

A

B

C

F

G

H

I

J

K

L

M

Q

R

S

T

U

V

W